The ORIGIN Of MOUNTAINS

The ORIGIN Of MOUNTAINS

By
John Delano
And
Thomas Oldfield

SECOND EDITION

Hopewell Junction, NY

Published by:

>John Delano
>
>38 Old Sylvan Lake Road
>
>Hopewell Junction, NY 12533

All rights reserved. No portion of this book may be reproduced or transmitted in whole or in part by any means, electronic or mechanical, including photocopying, or by any information storage and retrieval system without written permission from the publisher.

Copyright © 2008, 2015 by John Delano

Second Edition, first printing

Cover – The Eiger, a 13,000 foot dislocated mountain of the Bernese Alps carved by an ice flow.

Edited by Thomas S. R. Oldfield

ISBN 978-1-329-20312-9

Printed in the United States of America

This book is dedicated to
My children John, Chris, Paul and Lynn
For their invaluable inspiration
And my wife Connie
For her invaluable patience.

The ORIGIN Of MOUNTAINS

Theoretical Timeline of the Earth
(using most current methodology)

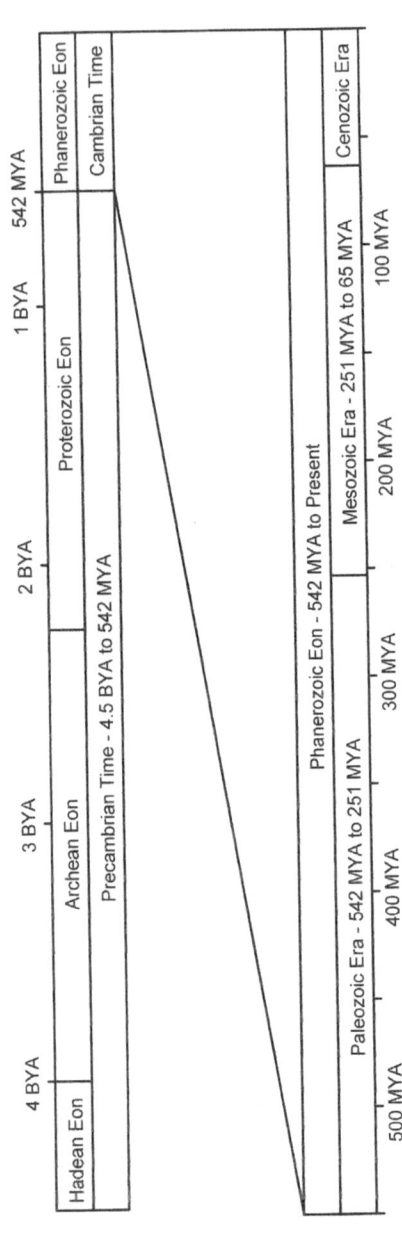

BYA   Billion Years Ago
MYA   Million Years Ago

copyright (c) Tier 7, 2014

Table of Contents

Preface 11

Introduction 15

Part 1   The Pacific Ocean Basin. 29
Part 2   The Eastern Flow 41
Part 3   The Indian Basin 49
Part 4   The Mechanics 57
Part 5   While the Ice moved 65

Conclusion 73

Afterword 77

Appendix 79

Preface

In 1974 the construction of the first of a new class of oil supertanker called the Very Large Crude Carrier, or VLCC, was completed by Seatrain Shipbuilding Corporation in Brooklyn, NY and delivered to the International Oil Tanking Company. It was named BROOKLYN, and it had three sister ships: the STUYVESANT, the WILLIAMSBURG and the BAY RIDGE. Together they would be capable of transporting a full six million barrels[1] of oil.

The months following this event saw the beginning of the Arab oil embargo, for the duration of which these ships, along with a new oil storage facility located on the island of Curacao called Bullenbaai, were fully stocked with nearly twenty-two million barrels of crude oil. This triad of producers, buyers and sellers was creating a shortage by the reduction in production and the stockpiling of the cargoes, thus increasing both demand and price. These VLCC tankers - among others - cruised offshore and were directed by their owners to travel where the shortages had resulted in highest prices.[2]

The world was in turmoil over the Arab oil embargo. Many towns in the United States saw altercations between customers at gas stations, and we all were upset that so sudden an event could so quickly throw our lives into disarray. Despite this, at that time America was not actually dependent upon the Persian Gulf oil, which constituted the greater part of the Arab supply. Seventy percent of the American demand was supplied internally

---

[1] 1barrel equals 42 gallons.

[2] The BROOKLYN class VLCC was rated at 225,000 tons with 1.5 million barrels of oil.

with the other thirty percent coming from Alaska, Mexico and Venezuela.³

In the late 70s economists stated that the world was moving toward an oil economy. I don't think many understood that this meant their livelihoods and activities would be dependent on those who controlled the production, shipment and refining of oil. Though given as a prediction, this was in effect an explanation of recent world events. The juggernaut was already in motion. The Seven Sisters dominated the oil industry and controlled eighty-five percent of the world's oil reserves.⁴

In Mexico, near the southern part of the Gulf, the discovery of a prolific oil well was in the newspapers. Why, I wondered, did some oil wells produce over 100,000 barrels of oil per day while others could not manage even a thousandth of that output? The first thing I considered was that oil wells - highly productive wells in particular - tended to be found in unusual places. When mapped, they appeared to fall along something approximating a ring, beginning with the north slope of the Brooks Range in Alaska, the front range of the Rockies, the state of Oklahoma, then Texas, the Bay of Campeche in Mexico, Venezuela, Nigeria in Africa, the Persian Gulf, and Russia's western Siberia.

---

³ The Alaskan oil was not considered internal because it was owned and controlled by British Petroleum.

⁴ The "Seven Sisters" was a term coined in the 1950s by businessman Enrico Mattei, then-head of the Italian state oil company Eni, to describe the seven oil companies which formed the "Consortium for Iran" cartel and dominated the global petroleum industry from the mid-1940s to the 1970s The group comprised Anglo-Persian Oil Company *(now BP)*; Gulf Oil, Standard Oil of California (SoCal); Texaco *(now Chevron)*; Royal Dutch Shell; Standard Oil of New Jersey (Esso) and Standard Oil Company of New York (Socony) *(now ExxonMobil)*.

* "Business: The Seven Sisters Still Rule". Time. 11 September 1978. Retrieved 24 October 2010.

* "MILESTONES: 1921-1936, The 1928 Red Line Agreement". *US Department of State*. Retrieved 18 August 2012.

* "Documentary: The Secret of the Seven Sisters". Retrieved 4 May 2013.

The Circle of Prolific Oil Wells

The next thing I did was purchase a globe. Due to the distortions of two-dimensional maps, I could be sure of this appearance of a circle by plotting the various wells on a three-dimensional surface. Looking at the globe I found a ring of prolific wells, those gushing over 100,000 barrels of oil per day. They did indeed fall on the edge of a circle and had as their center a point halfway between Iceland and Great Britain. Opposite this "oil circle" is the Pacific Ocean covering one third of the planet. These observations lead to further investigation, initially to determine the nature and location of oil deposits, but eventually to discover the cause of the present day geographic landscape of the surface of the Earth.

The currently accepted theory describing the placement of the continents is Alfred Wegener's Theory of Continental Drift, which he described in *Die Entstehung der Kontinente und Ozeane (The Origin of Continents and Oceans)*, first published in 1915. This theory gained further support in the early 1960s when Harry Hess of Princeton University proposed the idea of sea floor spreading.

With the data gathered by seismometers in the 1960s, the theory of plate tectonics was developed as the mechanism behind seafloor spreading and continental drift. These theories received further support from data gathered over the last two decades, regarding the ages of mountain ranges and the comparison of the fossil record on both sides of the Atlantic.

As part of my research, I studied this data, but arrived at a conclusion which differs from that described by plate tectonics, and particularly by continental drift. While I allow that the Earth's lithosphere moves and is divided into individual plates, I do not agree that the evidence points to continents moving great distances relative to each other. Along with this, I have developed an alternate theory explaining the formation of the mid-ocean ridges, without the mechanism of seafloor spreading and one which excludes them as the cause of plate movement or formation of mountains. In the following chapters, I will describe this new theory, which is sufficient to explain the topography of present day Earth without the need for continental drift.

John De Lano, 2015

Introduction

A very long time ago there were no mountains.

Most mountains that we see on planet Earth today are relatively young, and all mountain systems had a beginning and an end to their formation. The last great Fold Mountains, including the European Alps, the Rockies, Andes, and some ranges around the Pacific Rim were formed only a few million years ago. The greatest and largest of the world's mountains share the origin of its largest ocean - the Pacific, which covers almost a third of the earth's surface. The Pacific Basin was the starting point of the great mountain systems; a fact which can be deduced from physical features found on the Earth today.

As has occurred many times throughout history, the discovery of these submarine features was made by individuals engaged in a different pursuit altogether. In 1985, the US Navy had employed its newly launched GEOSAT satellite in the mapping of the height of sea level around the world. The sea height varied from the mean sea height as a result of differences in the density of the water depending on the depth of the sea floor. As a result, the same data produced by measurements of sea height could be used to map the ocean floors, including underwater mountain ranges and trenches alike. For three years, GEOSAT orbited the Earth. This was the first comprehensive mapping of the ocean basins, and it is only for the unexpected relationship between sea surface height and sea floor depth that we see the bottom of the oceans today as we never have before.

Two hundred and fifty million years ago, protozoa by the trillions, along with corals and crinoids, left behind the limestone we see today. We refer to this period as the Permian-Triassic

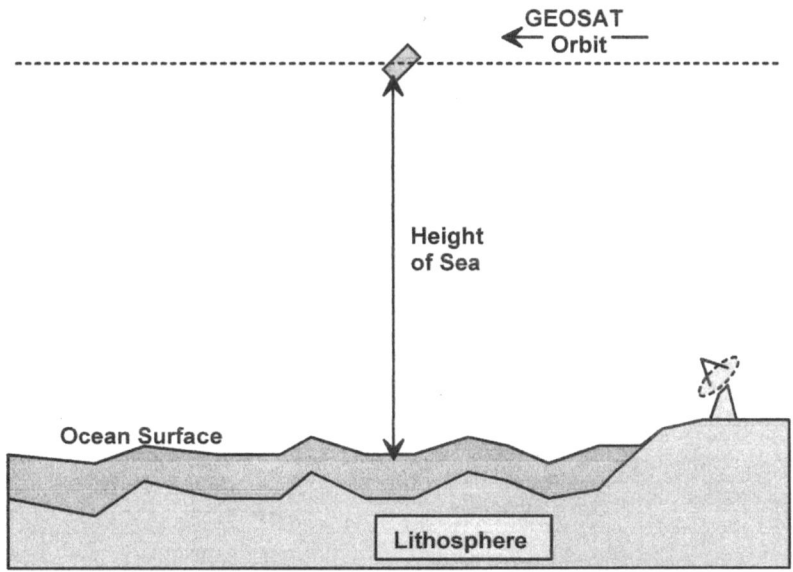

extinction event as the fossil record indicates that over ninety percent of marine species and seventy percent of terrestrial species disappeared at this time. But it is useful to look at it from a different perspective. The scientifically accepted cause of the lack of abundant fossil remains is the dying off of the major species, but this may not be supported by the evidence. The animals and plants did not entirely disappear from the planet, whereas the fossils did. I do not agree with the conclusion that a paucity of fossils implies that they were never laid down to begin with; instead, I contend that this layer of material was removed. There is evidence of an ice flow that scoured Brazil clear down to its bedrock and removed everything in its path - including the fossil remains – around 248Mya (million years ago).

    Louis Agassiz, the noted geological historian, was correct in his conclusions concerning ice marks on the vast plains of the Amazon in Brazil. He was the first to scientifically propose that the Earth had been subject to an ice age and published his findings in *Etudes sur les Glaciers* in 1940 and later his theory in *Systeme glaciare* in 1847. Agassiz observed glacial debris on the equator, but he and his work were marginalized by his colleagues, due to his staunch opposition to Darwinism.

Contemporaries of Agassiz - Goethe, Charpentier and Schimper had already concluded that the erratic[5] blocks of alpine rocks scattered over the slopes and summits of the Jura Mountains had been moved there by glaciers. As the Jura Mountains are over 5600 feet high, the glacier that would have deposited these rocks would likely have been miles high and would have covered most or all of Europe.

Vincent Courtillot, a French engineer, was critical of Giant Impact hypothesis[6] as a cause for mass extinctions and instead tied them to the ages of the flood basalts.[7] The age of the flood basalts and the dates of the mass extinctions coincide with a 97% probability.[8]

A flood basalt is created as a result of a volcanic eruption or series of eruptions that coats large areas with a solid layer of basalt lava. Flood basalt provinces are often called traps[9], which derives from the characteristic stair step geomorphology of many associated landscapes. The age distributions of twelve individual flood basalt episodes around the globe from the last 250Myr suggest that in most cases these took place over a quasi-periodicity[10] of 23Myr. The correlation with mass extinction is quite straightforward.

I suggest that the glacier that covered the Jura Mountains was merely the edge of a massive ice flow that destroyed all the fossil remains deposited over hundreds of millions of years. While the front of the flow removed fossils and crust alike, the ice would thin out at the rear of the flow, subjecting the crust beneath to

---

[5] An erratic is a rock that differs from the surrounding rock and has been carried there by a glacier from a distance.

[6] The giant impact hypothesis states that the Moon was formed out of debris from an indirect collision between the Earth and an astronomical body the size of Mars, approximately 4.5 billion years ago,

[7] "Mass extinctions in the last 300 million years: One impact and seven flood basalts?" 1994 *Israel Journal of Earth Sciences* 43: 255–266

[8] Appendix – Episodes of Flood Basalts

[9] From the Swedish "trapp" which means stairway.

[10] Hang one pendulum onto the end of another pendulum and you get a Quasi-periodicity, the erratic movement of the second pendulum.

lower pressures than those nearer the front. This permitted magma to escape from the mantle here, where most of the crust had been removed, and "ooze" out as molten lava.

The land, made level by the passing ice, would allow the lava to flow for hundreds of miles. The lava would attain a height of forty-five to sixty feet and harden. When another lava flow occurred and added to the top of the previous flow it would give the impression of "steps". One of the largest formations of this type is the Deccan Traps of India.

In terms of time, between 23 – 24Myr passed between the leading and trailing edge of the flow passing over the same point. The trailing edge may have been as much as three-thousand miles wide and would release the aforementioned stored magma along its entire breadth. These magma flows were influenced by the progression of the massive ice flow. In turn, the movement of the ice flow was chiefly governed by the movement of the Earth's poles changing the location of the equator and the 413,000-year (400 kyr-Eccentricity) large elliptical Earth orbit.[11] The 413kyr elliptical earth orbit caused tremendous increases in the rotational centrifugal force when it was in phase with other Earth cycles, and would cause the massive ice flow to rift. This sideway splitting from top to bottom of the ice shell was caused by its inability to push the six miles of earth forward along the shell's 3,000 mile edge at a rate greater than six inches per year and the side pressure caused by debris walls. This sideways rifting created the physical marks called transcurrent faults[12] and left linear imprints of sideways dragged bottom flows of liquid magma, as well as striping visible in the magnetic polarity of the basaltic rock, giving the impression of sea floor spreading.

It appears that the quantity of liquid magma present beneath the ice flow was inversely proportionate to the distance from the ice flow's center. These smaller volumes of liquid further out would harden the quickest. The conclusions drawn by Wegener and du Toit, two early continental drift theorists, from

---

[11] as calculated by Milankovitch. (Milankovitch, Milutin (1998) [1941]. *Canon of Insolation and the Ice Age Problem*. Belgrade: Zavod za Udžbenike i Nastavna Sredstva. ISBN 86-17-06619-9)

[12] Transcurrent faults cut across the main mid-ocean ridge at a 90 degree angle.

their observations were incorrect so the sketches they made provided erroneous information. The sketches gave the ocean floor, the physical transcurrent fault's movement based on the dates that the magma hardened. The fast hardening smaller volume of magma as you moved toward the edges hardened first. It did not move.

To illustrate, consider the following. There are three apple pies: one is 18" in diameter, the second a 9", and the third 4". Each pie is removed from their separate ovens, all baked to perfection with the filling and brown sugar bubbling and oozing from each pan. These are then placed on a windowsill to cool. After two minutes, the small pie is already cool, but the other two are still oozing. Four minutes later the 9" pie is cool, and only the filling of the large pie has not yet solidified. The question, however, is not at what rate did the pies cool, but instead how far did each pie travel while it was doing so. The answer is clear: after they were set onto the sill, they moved no further.

In depictions of the hot spots associated with the Hawaiian island chain, a small island to the northwest, Oahu, cooled thirty million years ago. Maui, a larger island closer to Hawaii but along the same ridge of the Emperor Seamounts, cooled only ten million years ago, but the big island of Hawaii has not yet cooled and still oozes lava. The intent of these depictions is to show that the two smaller islands to the northwest have traveled over the same hot spot, over which the island of Hawaii sits today.

There is not one hot spot over which the islands have moved. The Hawaiian Islands are the exposed tops of the Hawaiian Ridge, a thirty-six hundred mile range of over eighty underwater volcanoes. The reason that there is a larger volume of magma exuded by the island of Hawaii is due to the thinner, weaker crust in that location. This was caused by a massive Pacific Ocean ice cap ten to fifteen miles high, and six miles into the crust. The Emperor Seamounts to the northwest show the turning and the direction of the Pacific ice cap whose movement left the rebound footprint.

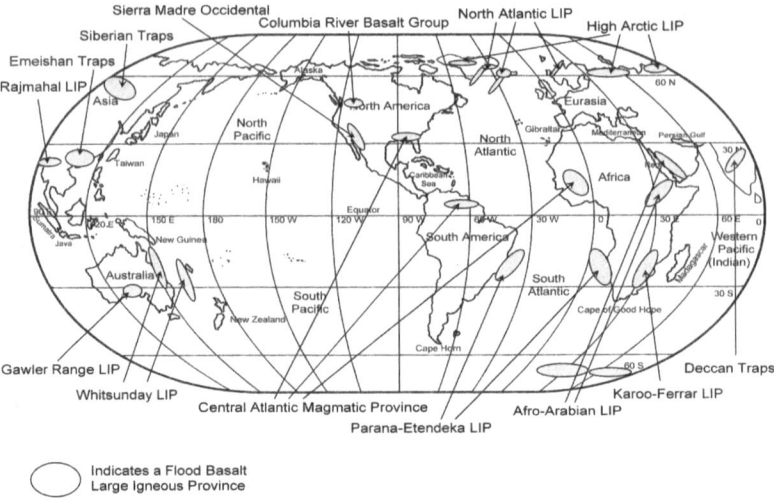

Large Igneous Provinces – Flood Basalts

Touching on flood basalts again, this would be a constant event and the intermittent nature of the number of flood basalts - twelve at last in the last 250Myr - is really in error. The event is continuous. The gaps appear to be a normal amount of erosion, about half of the lava over the eons.

The large igneous provinces, sometimes called false mountains, are so vast that if erosion removed half of the lava, the remaining twelve flows from the last 250 million years seems to leave about the right proportion remaining that we can see today in the trap regions. The footprints of man in the hardened lava flows of Hawaii tell us that the lava outpourings were not entirely dangerous and not likely the principle cause of mass extinctions.

Consider Lazarus plants, which can reappear millions of years after their apparent extinction. These are plants that were carried off by the ice flow in a dormant state and reestablished themselves in new areas as the ice receded. Consider the bipolarity of species – two ecosystems, isolated at the opposite poles of the earth, yet containing hundreds of genetically identical organisms. These phenomena, characteristically associated with the movement of continents, are tied to the existence of the flow of ice, and specifically to a habitable zone or narrow temperate region that existed on the equator.

Joining all the continents at the South Pole as suggested by the theory of continental drift, simply on the basis of ice markings and the theoretical directions of continental drift is perhaps untenable. Once you understand the ice flow's 248Myr circumnavigation of the earth, traced both by its footprints in the ocean floor and the rebound structures it left behind, you will be better equipped to see the vestiges of this journey where it crossed the continents themselves.

Global sea level fluctuations, specifically an increase which occurred from 250 to 65Mya, tell an interesting story. During a glacial age, sea levels fall as water is stored in continent-wide strata of ice; this trend reverses itself during interglacial periods. While the earth regularly experiences glacial and interglacial periods, the epoch of the ice shell and ice flows would have been characterized by significantly different climatological and geological phenomena, setting it as an outlier in this regard.

The Mohorovicic discontinuity, or Moho, is the boundary between the Earth's crust and mantle. The Moho lies almost entirely within the lithosphere; only beneath mid-ocean ridges does it define the lithosphere–asthenosphere boundary. This boundary appears to have been influenced by the movement of a massive ice flow and is thinnest under the mid-ocean ridges. Knowing that ice only has one third the mass of rock, the height of the ice shell can be estimated by the manner in which it affected the change in the level of land six miles deep. An ice shell many miles high over the modern Pacific Ocean moving, spreading and melting slowly over the Atlantic and Indian oceans over a period of 200 million years would cause a significant rise in sea-level in the oceans crossed by the equatorial habitable zone. This high water came from a voluminous melting of the ice shell about 65 million years ago. The Cretaceous–Paleogene extinction event, which comprised of the drowning of most land animals which lived during this period, was caused then not by a catastrophic meteor impact but by the melting of the ice shell. Despite the severe decline in species diversity, many sea creatures and dinosaurs survived.

The rising water would drive most dinosaurs to the high plateaus where the vegetation would be unsuitable for their diet, resulting in widespread starvation. Sea creatures may have suffered due to the change in the salinity of the oceans and species that moved into areas of shallow water would have perished when the high water retreated.

The Rocky Mountains of North America are Fold Mountains, and as with all other major mountain ranges, the process of their formation is not very well understood. Though Plate tectonics attempts to explain their formation as a result of the collisions between plates, many Fold Mountain ranges are not on plate boundaries. In contrast, this type of mountain range is frequently proximate to an ocean. The mountains on each side of the Atlantic Ocean were forming around the same time, but the plate borders are along the Mid-Atlantic Ridge, not along the mountains.

Most Fold Mountains have multiple parallel ranges inland, each which may differ from the other by millions of years in age. It has been theorized that these were caused by stresses of continuous collisions between land masses. This cannot account, however, for the existence of the Urals, Appalachians, Brooks and Kjolen Ranges, none of which lay along a plate boundary.

Another feature of the Earth which does not have an apparent explanation under the theory of plate tectonics is the Pacific Ocean. Covering almost one third of the Earth, it is not clear how this huge basin came to be, nor whether it has any relationship with the processes involved in the formation of the mountains along its boundaries.

In many research centers like those belonging to the US Geological Survey, Woods Hole Oceanographic Institute, National Oceanographic and Atmospheric Administration, Center for Lunar Origin and Evolution, and the Office of Naval Research, powerful supercomputers are working on the problem of how the Pacific Ocean basin formed and how the Earth acquired a moon. Scientists have established a probable sequence of events, a depiction of which which can be seen at an exhibit at the American Museum of Natural History in New York.

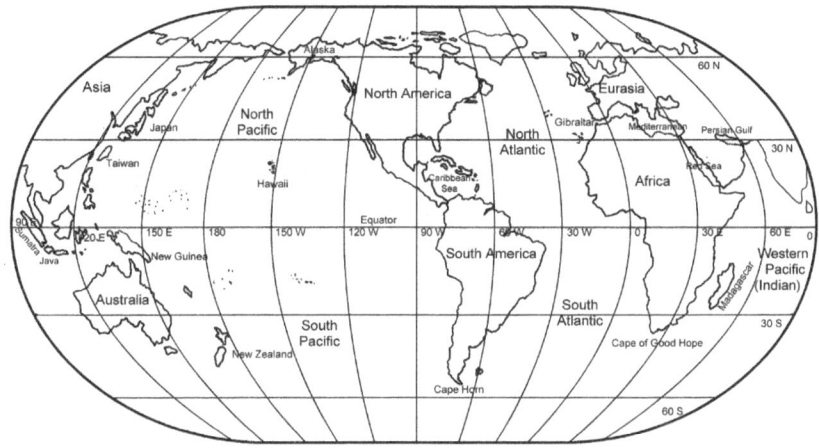

This exhibit, *"Our Moon"*, shows a dramatic moon forming collision between the Earth and another astronomical body, resulting at the point of the collision in the Pacific basin. This is called the Giant Impact Hypothesis and was proposed by Reginald Aldworth Daly of Harvard University in 1946.

First, it would be useful to discuss some facts about mountains so the importance of what are called fold mountain systems is made clearer. There are several mountain types, categorized chiefly by the way in which they were formed. A plateau is one such type, and is formed by natural forces acting to erode a larger, preceding mountain. When eroded even further, these eventually become dissected plateaus. The mountains of the Catskill Mountain chain, located in the state of New York in North America, are one example of a plateau mountain type.

Fault-block mountains result from rifting, which is where forces act on the crust to pull it in opposite directions. An example of a lifted fault-block mountain is the East African Rift. Tilted type fault-block mountains can be found in the Basin and Range area in the western United States.

Fold mountains differ significantly from dissected plateaus and fault-block mountains. Found on almost every continent, this formidable class of mountain includes the Rockies and Appalachians of North America, the Andes of South America, and the Himalayas in Asia, in addition to most major mountain systems of the eastern hemisphere. All, without exception, were pushed up by a horizontal force which originated from the same direction as the nearest ocean.

Each of the fold mountain systems have different ages of initial formation, and since no less than three continent size ice flows and a very high Pacific ice cap were the cause, it is reasonable to expect that these mountains would not only differ in their date of formation but that the dating would be serial for the flow involved. That is, we can follow the location of the ice flow by the age of the mountains it formed. There is a straightforward timeline.

The massive folded mountains that we see today, along with petroleum and coal deposits, were created over a period of about 600 million years. Even with this knowledge it is difficult to devise a system which accurately predicts the positions of such oil deposits. Exxon Corporation, for example, even with its 125 year history and experience still contends with a ninety-five percent failure rate for "wildcat wells" - those drilled for exploration outside known oil fields. This is on account of the formation of *petroleum* - the correct name for oil that formed during this period – still being not well understood. While the major fold mountain movements are related to the location of sediment basins - the drainage areas and outlet deltas – and these all appear to point to petroleum deposits, the exact nature of the process remains somewhat enigmatic.

It is easier to explain the origin and present location of coal, as we understand the sequence of its formation. Prodigious quantities of plant matter are required. A sufficient accumulation of biomass can form a peat bog which, when subjected to intense pressure and heat, undergoes the chemical transformation into mineral coal. We can look at a fragment of coal and see the imprint of leaves that went into its formation. Even given the timescales at work, the origin of the truly enormous volume of forest material required to produce the observed coal deposits requires an explanation. The forests must have been destroyed and regrown every hundred years or so, if we are to account for the quantity of the coal deposits observable today. It was the formation of fold mountains that supplied the necessary heat and pressure to form these deposits.

The theory of Plate Tectonics[13] is based on the theory of continental drift[14] in the lithosphere. It's more of an after the fact explanation than a tool for predictions. We've defined the plates by observing volcanic activity, earthquakes and seismic anomalies but knowledge of the plates has given us little information for the prediction of future events. The theory also does not address the driving force that moved the continents in the first place.

A rise in mountains today is more than likely caused by volcanic activity or the erosion of the mountains and the rebound upward by the reaction to the weight being lifted. With less overburden - that is, weight - on the lithosphere the mountain rises as the mantle rebounds. This rebound action is called "isostasy". One of the best examples of this is in Scandinavia where the mountains are still rising from the melting of the ice load from that last ice age.

From where did the ice and water on the earth come? In the billions of years preceding the formation of the first mountains, the cooling rock of a newly-formed earth off-gassed steam that became water. During this same period comets, the mass of which is around eighty-five percent water, bombarded the Earth. These collisions with Earth do not appear to have left marks on the surface. If they did, some subsequent mechanism has removed them.

They may have struck a thick layer of frozen ice, in which case the impacts would not have sufficient energy to penetrate all the way through to the crust. The effect of tectonic plate movement may have removed any signs of early impacts as well. It is also possible that a continent-sized ice flow moving across the surface could have erased all signs of such early bombardments.

---

[13] General theory formed between 1950 - 1970

[14] Abraham Ortelius – Thesaurus Gepgraphicus, 1596; Alfred Wegener, 1912

During the Nixon presidency we visited the moon. Using radiometric dating techniques we determined the age of the moon rocks that were brought back to be about four and a half billion years old. It was, then, four and a half billion years ago that Reginald Aldworth Daly's theoretical moon forming collision took place. Though most subsequent impacts have been erased - by one force or another - from the earth, the moon has preserved the record of such collisions on its surface.

The moon's small size and lack of an atmosphere made the moon incapable of retaining large quantities of liquid water, and therefore ice did not form in thick layers to cushion the surface from the perpetual bombardment to which it has been subjected over the eons. The moon's inception, as calculated by computer models and our study of moon rocks, can be theoretically attributed to an ancient collision between earth and another protoplanetary body. The resultant material that would have been thrown into orbit by the collision is believed to have formed into our moon.

The moon is slowly drifting further away from Earth[15], due to the effect of Earth's tides slowing down the Earth's rotation. If we go back in time, with larger ocean's tides and figure where the moon should have been four and a half billion years ago it would show the moon almost within the theoretical "Roche Limit".[16]

In this work the theory that we will examine is not only useful in explaining the formation of today's mountain systems, but can also be used to predict the location of petroleum and mineral deposits. It will lead to the understanding of why fossils are found in specific strata. With one pivot point on land in South Africa it establishes that continent as the beginning point of all the plants, and animal's migration routes over the past 124

---

[15] Determined by the Lunar Laser Ranging Experiment to be 1.5 inches per year.

[16] According to the theory proposed by Edouard Roche in 1848, planetary objects are held together by gravity, and if one gets too close to another larger object held together by gravity, the smaller object breaks apart. The debris may then revolve around the larger object in rings. Roche Limit Calculations can be found at http://cs.mcgill.ca/~rwest/wikispeedia/wpcd/wp/r/Roche_limit.htm

million years. It also explains outliers, such as the Lydekker Line and the strange creatures of Australia.

Earth had to be very different one and a half billion years ago. I don't believe our planet was covered with water at that time; I think it was covered by ice. This work describes the development process of the Earth if this indeed was so. I propose it as the Ice Flow theory.

Part 1

THE PACIFIC OCEAN BASIN

We might expect to find the surface of a planet relatively smooth. This is the intuitive expectation, as even with geological activity raising mountains and astronomical activity creating craters, atmospheric activity will concurrently wear down these features, leaving only faint evidence of their existence behind. The Earth's surface does not look like the surface of any other planet in our solar system. The crust of the Earth is very rugged with deep depressions and trenches and high plateaus and mountain ridges. Water covers many of these features, including most of the volcanoes, and though we have some evidence of craters the examples are not prolific. Our major mountain systems, however, are quite different from those of our neighbors.

A moon forming collision as proposed in the Giant Impact Hypothesis would have left a large depression in the Earth's crust, both from the depression of the compacted material of the impact itself and from the material lost by the impact. There appears to be such a depression, covering one third of the Earth's surface and containing half of the Earth's seawater, called the Pacific Ocean Basin. When we look at this ocean's floor, we find an abundance of volcanic activity; in fact, there is an order of magnitude greater activity in the central and western Pacific than anywhere else on Earth.

Another interesting geographic feature of the Pacific is that it is bounded on all sides by fold mountain ranges: the Koryak and Alaskan Ranges to the north, the Rockies and Andes to the east, the Transantarctic Mountains to the south and the Great Dividing Range and Sikhote-Alin Range to the West. To the north and east, tectonic plate boundaries coincide with the mountains and we may by observation draw the conclusion that the former is responsible for the latter. This would be a hasty and inaccurate conclusion to draw, however, as to the south and west there are mountains without a proximate plate boundary to have created them. There must have been some other action lifting the crust. The giant impact is another possible force which could have pushed up these ranges, but this too is incorrect. We also have fold mountains on the shores of the Atlantic, where there are neither plate boundaries nor known sites of significant meteoric impact to bring such landforms about.

On observation, the floor of the Pacific Basin appears to be scratched and gouged. This seems to indicate that something hard scrubbed or plowed around in this basin and moved the debris out from the center, pushing, piling and folding it up on our continents of today. There is trace evidence in this scarring which indicates a clockwise rotation around the basin, leaving a linear non-random pattern showing the direction and speed of the plow.

There is evidence of a massive ice cap or shell which formed over the surface of the Earth, thickest at a point centered in the Pacific region. The north polar region may have been located in the area of the central Pacific today with the south polar region in the area of the south central African continent. Based on the comparative density between rock and ice and the deformity left in the Lithosphere, this shell would have had a height of between fifteen and twenty miles.[17]

---

[17] This is based on the depth of the basin, the difference in density of rock and water, the volume of water on Earth today and the water lost through 4 billion years as identified by Pope, Rosing and Bird in their 2010 study for the Natural History Museum of Denmark.

Footprints of Ice Movement

The ice at the bottom of the cap acted as an insulator trapping the heat of the Earth's core, and the heat from the decay of radioactive material in the crust and mantle. The ice cap was an asymmetrical mass on the surface and it caused three things to happen. First it caused the Earth to wobble.[18] Next, the wobble caused the ice cap to move. Lastly, when the ice cap moved, it started pushing the crust toward the equator, generating heat via the friction between the bottom of the ice cap, the shearing crust and the mantle.

Under the considerable stresses applied by this inner heat and the centrifugal force of the rotating Earth, the ice cap separated and moved in two directions. One part flowed up through the present day North Pole toward central Russia and the other turned clockwise in the Pacific Basin. I'll deal with the movement toward Russia in the next part of this work. The

---

[18] This is called the Chandler Wobble named for Seth Carlo Chandler who identified it in 1891. The magnitude is approximately 30 feet from the axis.

clockwise motion in the Pacific was due to the Coriolis Effect caused by the Earth's rotation.

The moon's low density and lack of an iron core support the Giant Impact theory's assertion of consolidated debris making up the moon. However, radiometric dating of the Pacific floor puts its age at about two hundred million years, which instead supports the Ice Cap hypothesis.

Consider that the moon forming impact not only melted the ice across the entire planet, but created a hollow or basin where the Pacific Ocean is today. The water and vapor that was released from the cooling crust, along with that deposited by the impact of comets, would have filled in this "Pacific hollow".

As the Earth cooled the water would have turned to ice. With a thinner atmosphere, warmed by a sun which was about 9% cooler than it is today, the temperature on the surface would have been well below freezing. For three billion years the earth would remain in under these frigid conditions. This volume of the ice in the Pacific hollow would increase due to the expansion of the frozen water, and rise high above the surface of the surrounding continents.

Study of the "snowball earth" hypotheses by Joe Kirschvink has uncovered evidence of a layer of ice extending down to the equator during the period of 850 to 500 Mya. This frozen earth period may have started before this by a billion years or more. That there was an open ocean during this latter snowball earth only on the equator is one conclusion I take away from these studies.

The basin of the Pacific contains fractures and deep trenches in the central and western areas with a congestion of volcanic activity leaving the area with thousands of islands, reefs and shallows amid deep water. The northern, eastern and southern areas have fractures and trenches at the borders with an open central plain that is marked by a continuous series of ocean ridges.

Looking at a chart of the Pacific basin one can see the ridges to the east. The part of the ice shell that remained in the Pacific pivoted in a clockwise direction around a point in the vicinity of where the Hawaiian Islands can be found today, with the ice flow pushing up the fold mountains on the west coast of the Americas as it moved south toward Antarctica. The age of these mountains gives us a good reference for direction; the Alaskan range has been dated to 150Mya, the Coast Mountains of Canada to 130Mya, the Rockies from 120 - 40Mya and the Andes to 25Mya.

The ice remaining in the Pacific was plowing the crust outward from the pole, toward the equator. The Earth spins on an axis through the north and south poles but we would be mistaken to assume the axis has always been where it is today. As the Earth's axis shifted toward its present location the equator also shifted. The ice continually flowed toward it so the direction of the ice flow changed.

Fold mountains, when you fly over them, resemble the waves in a stormy ocean. In order for the rock of these mountains to fold in such a fashion, there would need to be a force capable of acting on entire mountain chains, and over sufficient timescales that the naturally rigid basalt and granite express sufficient ductility to deform instead of fracture.

As an example, consider a large parking lot after a night of heavy snow. Everything is level until the local pickup truck arrives with a snow plow. It begins at one end and clears the snow by pushing it out of the way of the traffic until he reaches the other end of the lot. He will create "folded snow piles", with some piles looking like waves in the ocean. Some piles will be older than other piles, and it may even be possible, by careful examination of the density and thickness of the folds or texture of the ice, to determine which piles were formed first, as well as the direction and speed at which the truck was traveling in creating them.

While we may expect to see depressions where the heaviest ice rested and moved, we instead see the opposite in the form of the mid-ocean ridges. These ridges are actually rebound structures.

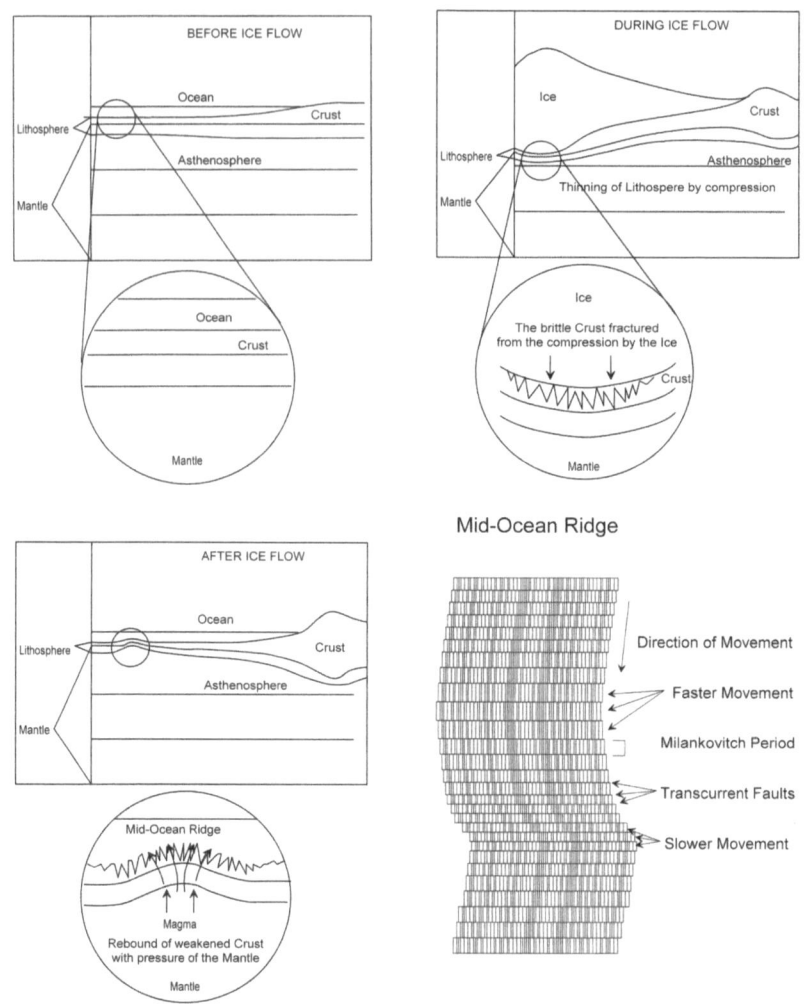

The Development of a Mid-Ocean Ridge

The ice scoured the ocean bottom, removing most of the crust. The scraping indicates the direction of movement. This also explains why these parts of the oceanic crust have an age around 200Mya; the older material was removed leaving a weakened bottom compressed and thinned which allowed magma to ooze out from the upper mantle.

In the 18th century, the geologist Sir Charles Lyell posited that the earth was very old and that the sedimentary layers would be older as one dug deeper in the ground. He likewise expected to find proportionately older fossils in these deeper layers. This was an era before the modern methods of dating of rocks by way of radioactive decay; all strata or layers of rock were dated by the fossils found in them. This became known as modern uniformitarian geology.

Today, most earthquakes are due to the rebounding effect of this early movement of ice. I call these ridges rebound structures because the central weight of the ice flow was in the middle of the ridge, and here the crust experienced the greatest compression. As the ice melted, the crust and the mantle below began to return to equilibrium, and the area of the ridges rebounded the most. As the lithosphere had cooled and hardened, the surface fractured, exposing magma and leaving a weakened, striated surface. One example of this would be Iceland which, on account of this rebound effect, has been rising from the sea bottom for 200 million years. The ridge appearance has been removed by erosion, the action of the sea and continuous volcanic activity.

As an exercise, imagine the following: take a tabletop globe and rotate it so that you can see the Pacific Ocean. Place the middle finger of your left hand on Mid-Atlantic Ridge, and the middle finger of the right on the East Pacific Rise, letting your thumbs meet in the Pacific. Now, follow the ridges with your fingers, letting your thumbs move one degree of latitude north for every one degree of latitude your fingers move south. By doing so, you can visualize the movement of the Pacific ice cap.

When we discuss the origin of mountains we must take into consideration why in some places rock layers or strata are missing. Even more importantly we must attempt to explain why some strata are chronologically reversed, overturned, or even upside down. For example, the Matterhorn, a well-known mountain in Switzerland, is made of old gneiss, but the layers beneath are younger felsic intrusion. Some geologists say this mountain was uplifted and pushed horizontally cross-country for thirty to sixty miles. It has been hypothesized that this action was

The Lewis Overthrust

a result of tectonic plate subduction, but this is unlikely as the same condition would exist throughout the region. In that theory, the Alps would rise together, yet the top of the Matterhorn is African, not Eurasian.[19] Other geologists believe the Matterhorn was placed in its present location by a glacier. Most believe the top was shaped by glacial action. At almost three miles high, the shape and location of the Matterhorn supports the theory of a continental size ice flow that moved in a northerly direction.

Another example is the Lewis Overthrust,[20] a displaced mountain. Located in Glacier National Park in northern Montana, the Lewis Overthrust is a thrust fault, giving it the appearance of being upside down. Oil drillers report that older rock sits on top of rock strata that are 500 million years younger.

---

[19] Michel Marthaler – The Alps and our Planet: the African Matterhorn, a geological story. 2006 National Library of Australia

[20] The document that describes this geology is The Lewis Thrust Fault and Related Structures in the Disturbed Belt, Northwestern Montana - Geological Survey Professional Paper 1174.

The Colorado Plateau, covering an area of 130,000 mi^2, is located at the intersection of four states. This plateau, pushed up from the west, drains to the Colorado River from a significant height with a tremendous potential energy that has eroded rock for millennia, creating what is called the "Grand Staircase".

The canyons created by the Colorado River include the Grand Canyon, Zion Canyon, and the formations of Bryce Canyon National Park, but there are many more.

Yellowstone National Park shows huge calderas - or depressions - that remain after the collapse of land around a volcano that has erupted. This happened after the Front Range of the Rockies was pushed from the west over this land. The same force that created the Colorado Plateau also created magma under the mass from the heat of friction and pressure. This huge mass of mountain building was moving above, and the magma source has been traced to the lands to the west under the mountains.

The huge Yellowstone calderas are old volcanic areas, and lava flows from them are the byproduct of the mountain formation hundreds of millions of years in the past. The Front Range of the Rocky Mountains were being pushed from the west by the Pacific ice cap, with smaller ice flows pushing the uplifted land. The mass was moving toward the ever-changing equator.

These massive movements of mountains were common in proximity to the Pacific ice cap and as a result were neither small nor localized. The Lewis Overthrust is a hundred and thirty miles long, thirty-five miles wide and three miles deep. This area seems to have been pushed thirty to forty miles until it came to rest on its current foundation of younger rock and soft clay, something which may seem hardly possible.

An ice flow could have moved the Matterhorn or created the Lewis Overthrust without destroying the comparatively fragile folds of either monolith or the softer base material. The advancing ice would have sheared off and encapsulated these areas before carrying them away.

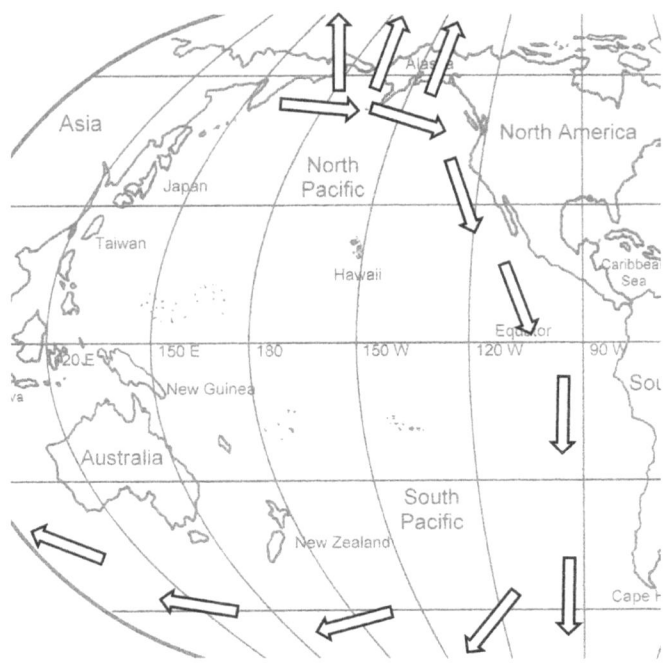

Ice Cap Breakup and Movement in the Pacific Basin

When you look at a topographical atlas, you can see thousands of miles of underwater mountains. These ridges are striated in two directions and indicate the movement and direction of the ice as well as the speed at which it moved. The mechanism that formed these 10,000 to 15,000 foot high mountains under the oceans started from the Pacific Ocean.

I regard to the speed of movement and the diagram of the development of the mid-ocean ridges, the speed indicators are the transcurrent fault lines and they line up chronologically with the Milankovitch Cycles.[21]

In the western Pacific the ice flows were influenced by their own centers of gravity, moving and clashing where they came together. Such collisions could cause the flows to merge or fold upon one another, generating colossal forces capable of gouging deep trenches like those around Fiji, the Marianas, and Japan. The

---

[21] Milutin Milankovitch, Serbian astrophysicist – calculated the different Earth cycles including precession, obliquity and eccentricity and the period for all cycles to coincide with a maximum instability of the Earth every 413,000 years.

eastern Pacific ice flow moved south from the Baja Peninsula. It pushed up the mountains of the Americas while the main body was channeled south to Antarctica, where it was redirected west. In doing so, Antarctica was subject to the full force of the flow, lifted by it and by result becoming highest continent by average elevation.

    The flow continued westward with the main body plowing through the lithosphere, separating Australia from Antarctica. It then headed into the Indian Ocean basin where it turned north at the Kerguelen Island platform, eventually meeting the Atlantic flow in the Rodrigues Fracture Zone. A second, smaller part of the flow, a mid-Pacific flow, turned briefly north, scouring trenches in the western Pacific, passing New Guinea and subsequently gouging out trenches on its way to the Philippine Sea.

Part 2

THE EASTERN FLOW

As noted above, when the Pacific ice cap separated, part remained in the Pacific and part moved toward northern Russia. Having described the movements of the ice in the Pacific, the following section will describe the path taken by the Russian flow.

As the ice moved over the present day Arctic region, the primary body of ice flowed toward central Asia and a secondary flow broke off to the right flowing between Greenland and Scandinavia. Just as the Pacific ice cap turned clockwise in the Pacific basin due to the spinning of the Earth on its axis, the eastern flow turned clockwise. The main body of the eastern flow's progress was blocked by the immoveable structure that it created, the northern part of the Tibetan Plateau, the 250-500 million year old Tian Shan[22]. The path is indicated by the Ural Mountains. This range is actually a rebound structure which is now above sea level. Like the ocean ridges, the entire Ural mountain chain in Russia is a rebound structure. The flow which created this chain would have been huge, coming over the present day arctic directly from the Pacific and forming the folded mountains along its path.

While the flow was powerful, the bulk of the crust driven in front of it eventually became so great as to be immovable. This bulk is now the high continent of Asia, which would block any further advance of the Russian ice flow. The secondary flow

---

[22] Tian Shan – (Heavenly Mountains), a 1900 mile east/west wall of mountains

which had no such obstacle and broke to the west, continued to move unimpeded towards the equator as the Atlantic ice flow.

The Russian section of the Ural Mountain chain, along with the entire Russian platform, was once at the bottom of a deep ocean. Driven by the radioactive decay of unstable isotopes, the material of the lithosphere beneath the Russian platform expanded. As the ice above disappeared, this expansion resulted in a steady upheaval of the platform itself, as much as three miles above its previous elevation. This has left Russia and the Ural Mountains in the middle of Asia above sea level. Lake Biakal in Russia is an ocean trench above sea level, as is the Great Rift Valley in Africa. Evidence for this sequence of events includes the rare metals mined in the Ural Mountains, such as gold and platinum, which are typically found in mid-ocean ridges.

An ocean ridge has been described as an underwater mountain system, typically having a valley known as a rift running along its spine, formed by plate tectonics.[23] This is an imperfect description, as some ocean ridges are not near plate boundaries. A more accurate definition would be 'a rising of magma which results in the formation of volcanic mountains along its entire length'. The magma pools that feed the center of the ridges are still oozing magma from below and most of it hasn't hardened yet. The edges of the hardened lava flows, still part of the ridge system, hardened over 141 million years ago as shown on the following diagram.[24]

In the Arctic region the secondary flow that separated from the primary body of ice can be tracked by the footprints as it turned to the right and flowed between Greenland and Scandinavia. This flow was of continental proportions and scoured portions of the Americas and Africa.

---

[23] Merriam-Webster – Mid-ocean Ridge; www.sciencedaily.com/articles/m/mid-**ocean_ridge**.htm; www.trinity.edu/gkroeger/.../notes/intro%20plate.htm

[24] John Wiley and Sons - 2008

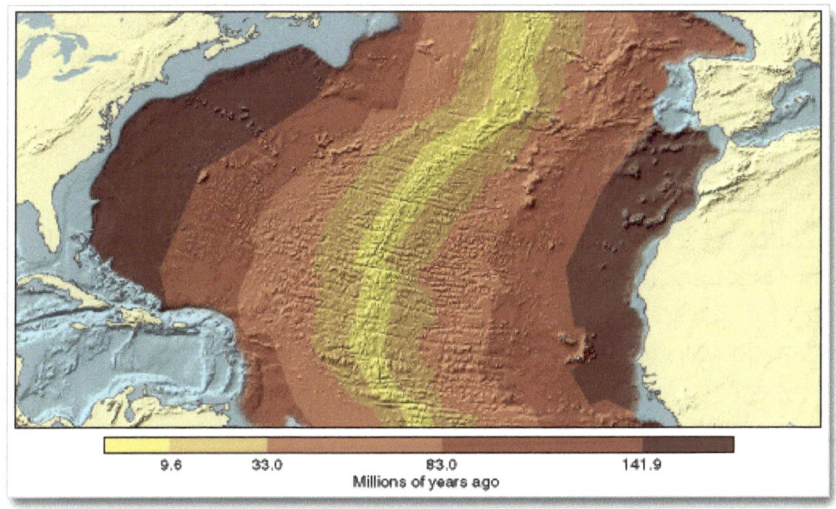

9.6   33.0   83.0   141.9
       Millions of years ago

    The Mid-Atlantic Ridge is one of the most interesting of the ridges and the second best defined.[25] This was thought to be a standalone feature in the Atlantic. This ridge is not, however, a single feature and it wasn't formed by a single event. The Mid-Atlantic Ridge is not limited to the North Atlantic, but is part of a worldwide system of ridges, created by massive volcanic upheavals, which encircle the globe. From the features of these ridges, one can determine the direction of the flow which formed them, and when it did so.

    Many geoscientists believe that seafloor spreading is taking place at the site of the Mid-Atlantic Ridge, and that this is the most convincing argument for the theory of plate tectonics.[26] The picture shows the result of testing performed on the Atlantic Basin. The rock tested nearest the ridge was found to be the youngest, growing proportionately older as its distance from the ridge increased. Seafloor spreading is not the only explanation for this phenomenon. Continental Drift theorists[27] believe South America and Africa were once joined, with seafloor spreading

---

[25] The best defined ridge is the Mid-Oceanic Ridge in the Indian Ocean.

[26] USGS - http://pubs.usgs.gov/gip/dynamic/developing.html; American Meteorological Society - http://www2.ametsoc.org/amsedu/online/oceaninfo/samplecourse/oceanchap2.pdf

[27] Understanding Earth, Grotzinger/Jordan; Ency. Britanica, Evolution of the Ocean

offering an explanation for their current separation. However, GPS measurement indicates that the distance between these two continents is decreasing, an event which is incompatible with the notion that the floor of the Atlantic Basin is expanding outward from the ridge.

The next two diagrams compare the movement as described by Continental Drift theorists with the results compiled from GPS data. The theoretical movement does not correlate very well with GPS data; South America, for example, is not moving in the direction predicted by plate tectonics.

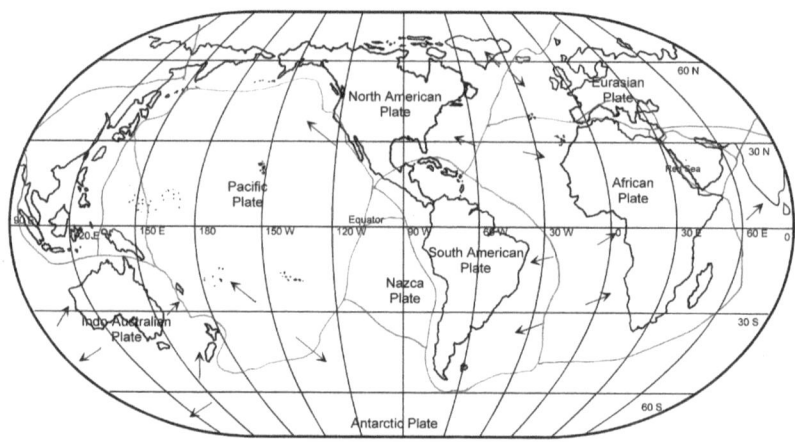

Plate Tectonics - Continental Drift

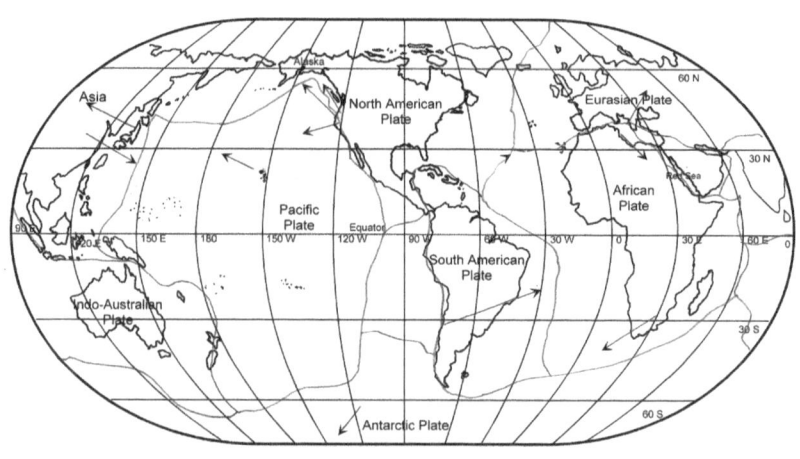

GPS Tracking

If the ocean floor melted from the heat generated by an ice flow or by magma flowing from the center of the ridge, the lava at the edges would be thinner with the reduction in volume and thus would solidify much earlier than that at the center.

Plate tectonics is not the answer. There seems to be a discrepancy between theoretical and real world geology. How can all these plates move with sufficient speed to create mountain ranges of different ages far from the plate boundaries like the Appalachians, Alps and Transantarctic? Why are seamounts found across the ocean floor by the thousands, yet so rarely appear along ocean trenches? Why is South America moving in the wrong direction?

Plate tectonics indicates that the Indian landform is said to have drifted on a plate 5,500 miles and pushed up the Himalayan Mountains and Plateau. However, the Indian plate is level at the foot of the Himalayas; it does not form a subduction boundary. According to Continental Drift, the ocean plates, on account of their greater density, are subducted, or sink, under continental plates.

Like the mountains of the Pacific, the mountains of the Atlantic are chronologically ordered, older in the north and younger in the south. The Mid-Atlantic Ridge begins in the Arctic. The Ural flow, after stopping at the Himalayan platform around 500Mya, spilled back over the Arctic and pushed through between Greenland and Scandinavia. It plowed up the northern Appalachians in America around 400 Mya and the Atlas Mountains of Africa around 200Mya, pushing up the southern Appalachians around 185Mya. At 150Mya it was pushing up the Guiana Highlands and Brazilian Highlands. The Karoo large igneous province[28] in southern Africa has been dated to 133Mya. This is all indicative of a large serial event in the Atlantic basin over 124Myr with a speed, as mentioned in the previous section, of six inches a year.

The main body of the ice flow, being very high, spun off secondary flows that moved ahead and to the sides of the central

---

[28] Thjis is an extremely large accumulation of igneous rocks, including liquid rock (intrusive) or volcanic rock formations (extrusive), when hot magma extrudes from inside the Earth and flows out.

flow. Given their elevated position, they did not need to overcome the resistance of the lithosphere. These secondary flows were different, however, as they would have crept along like large glaciers off the flow. These would have been responsible for such features as the Gulf of Mexico.

The reason the coasts of the Americas and those of Europe and Africa look roughly like pieces of a puzzle that fit together is that one ice flow carved them out at the same time. The mid-ocean ridges show us the path the ice took and the transcurrent faults show us the speed that it traveled at. These ridges are still rising today.

The Appalachians, a 1,500 mile chain of fold mountains running parallel with the east coast of North America, formed in a period no greater than 50 million years. The youngest of the Appalachian chain of mountains is located in Tennessee and Alabama. Interestingly, it has also been proposed that the Ouachita Mountains of Arkansas and Oklahoma may be a continuation of the Appalachians, which would make them its youngest members instead and support ice flow as a mechanism.

South Africa has fold mountains, appropriately called the Cape Fold Mountains, and huge false mountains or plateaus of flood basalts, such as the Karoo. These were created with a push that came from the west and then the south as the ice flow rounded it as indicated by the noticeable curving arc of the mountains around the southern end.

The Karoo LIP (large igneous province) has a particularly complex history. Under plate tectonics, this would constitute a flood basalt formed from the separation of Africa and Antarctica, but the rotation of the Karoo basalts during their early break up period, and the varied ages of the basalts which correlate with the 413,000 year Milankovitch cycle are more compatible with a large ice flow rotating around south Africa from the Atlantic to the Indian Ocean.

If viewed from satellite, the Karoo exhibits circular formations, long suspected to be the remnants of ancient impacts, but by the early 1990s, with satellite imaging, these circular landforms were highly suspected of being hot spots.

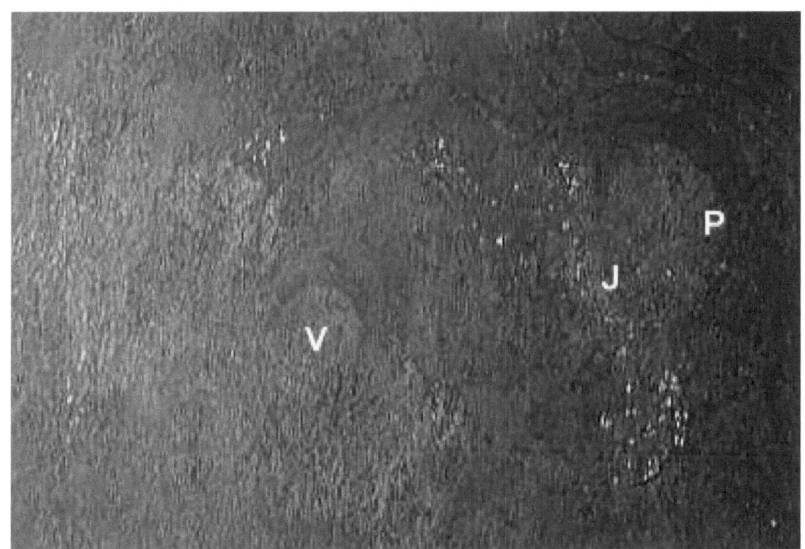

Circular Formations on the Karoo LIP

"Judged from cloud behavior, it is most probable that temperature differences exist between the circular areas and the surrounding land. On the satellite TV-image (June 1994) it was very clearly seen after a night with heavy frost, that the North-West circular area showed up dark against a white background owing to greater radiation from the circular area than from the surrounding land. These circular areas are thus apparently "hotspots"[29] that affect condensation of moisture and also atmospheric pressure and wind. [30]

The West Save Crater, located at 32°17'25"E, 20°03'25"S, was examined with a N-S ground magnetic traverse...during the 1993 Zimbabwe Impact Crater and Meteorite Expedition (Reimold et al., 1994). Detailed study of the craters showed the presence of mafic volcanic

---

[29] The term "hotspot" is also used by Plate Tectonics theorists as a heat source from the Earth's core. That is not the case here. This case refers to volcanic magma in the upper mantle.

[30] (THE OCCURRENCE OF CIRCULAR ANOMALIES IN THE KAROO AND SURROUNDING REGIONS, *P.W. Roux Grootfontein Agricultural Development Institute, Private Bag X529,Middelburg CP 5900,*Published Karoo Agric, Vol. 6, No 1, 1994 (3-5)

rock (basalt and gabbro) in the centre, surrounded by undeformed but indurated, and therefore more resistant, sandstone, which weathers out prominently, forming the ring-shaped rim around a central crater-like depression. These structures are interpreted as volcanic pipe-like feeders to the Karoo lavas which form the uppermost unit of the Karoo Supergroup (Duguid, 1978; Reimold et al., 1994). In the past they have also been investigated as possible kimberlite pipes." [31]

These circular formations are the remnants of volcanoes. An ice flow moved around South Africa 124 million years ago and in its course, it sheared off huge volcanoes leaving circular formations in place of their calderas. The volcanic process left gold deposits in the immediate region as it typically does. Diamonds found in the "pipes" or necks of extinct volcanoes are well-known in South Africa, and offers further proof of the true origins of the circular formations. More examples of the same circular remains can be found on the west coast of Africa in Mauritania, and diamonds are also found there.

The flow did not stop at the southern end of Africa but turned around it and headed northeast to the central Indian Ocean where it met with the Pacific flow in the Rodrigues Fracture Zone, as mentioned in the previous section.

A geological clock is frozen in the fold mountains of the world, and its center is located in the heart of Africa. The hand of our geological clock holds a secret, and that is that all fold mountains were created at the rate of six inches per year. At 12:00 lie the Altai Mountains in southern of Russia, 500 million years ago. At 11:00 the Caledonian Range in Scandinavia dates to 400Mya. At 9:00, the Appalachian mountains in North America date to 186 Myr; at 6:00, the Cape Fold mountains, South Africa, 124 million years old and at 3:00, the Himalayan mountains, India, 62 million years old. Lastly, at 1:00 and only 20 million years old, lay the young mountains called the Alps, in Europe.

---

[31] Distinguishing between impact craters and volcanic pipes using ground magnetics: field examples from Zimbabwe Sharad Master, David J. Robertson

Part 3

THE INDIAN BASIN

The first part of this book concluded with the main body of the Pacific ice flow moving west under Australia and a secondary flow coming north of New Guinea to the Philippine Sea, and the second part ended with the Atlantic flow moving east under the continent of Africa. The Indian basin is the area in which these two flows converge.

120 Mya the Atlantic ice flow was rounding South Africa, the southern Pacific flow was cutting Australia off from Antarctica and the mid-Pacific flow was turbulently overrunning the Coral Sea. Twenty million years later, the Atlantic flow was on the verge of splitting south of Madagascar, the southern Pacific flow was pushing up the Stirling mountains of south west Australia, and the mid-Pacific flow was spilling into the Banda Sea. Ten million years later still, they would be reunited for the first time since the Pacific ice cap split more than five hundred million years earlier.

The main body of the Atlantic flow and the southern Pacific flow collided at the triple junction in the Rodrigues Fracture Zone while the mid-Pacific flow flooded into the eastern Indian Basin. The combined flows were in the central Indian basin, applying pressure to the mantle and spreading as the mid-Pacific flow moved west. The rebound structure that resulted from this joining of ice flows is the most dramatic example of its kind.

The newly conjoined Indian flow and the mid-Pacific flow met at the Ninety East Ridge in the east central Indian basin. Though the Ninety East Ridge looks similar to the mid-ocean ridges, it is not the same type structure. This is not a rebound structure with a rift valley; this ridge was formed by the plowing action of the conjoined Indian flow and the mid-Pacific flow.

Over the next fifty-five million years the two flows moved fifty-five hundred miles toward India into the Bay of Bengal and the Arabian Sea, still at the consistent six inches per year. Secondary flows dug out the Rift Valley in Africa, the Persian Gulf, the Bay of Bengal and the trench in the Timor Sea which separated the fauna of Indonesia and Australia in what we define today as the Lydekker Line.

In the diagram, a line of interconnected red dots shows the distance along the Sunda Trench from 5,000 miles to 500 miles in the north. It shows the western side of the trench moving four times faster and the in the opposite direction of the "Sundaland" to the east.

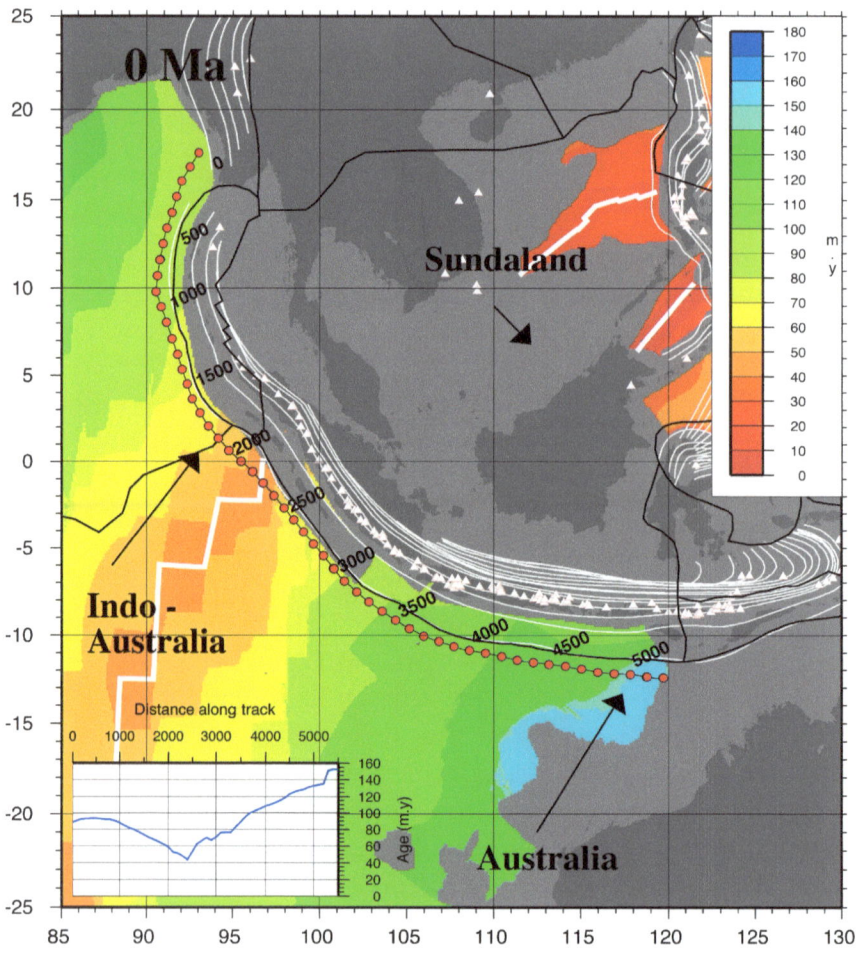

Notice the trench offshore and arc-like shape of the island nation of Indonesia. This is an example of an island arc, which is formed at the boundary of two ice flows moving in the same direction. The dominant Indian Ocean ice flow created the trenches and lifted the land by creating deep magma pools under the land ward side of the arc. In this area, the fault in the lithosphere or crust is the Sunda Trench. The ocean side of the trench in the north moved northeast, and though the land under Indonesia moved in the same direction, it moved at a slower rate.

The magma pools, reacting to a series of large earthquakes, triggered a mass of volcanic activity along the edge of this island arc. This is why this still-active collection of volcanoes, represented in the chart by small white triangles, are all around the same age.

The fold mountains made by the single Atlantic ice flow were diminutive compared to those made by the combined force of the two ice flows that came together at the Indian Ocean.

The Kerguelen Plateau is a submerged subcontinent 250 miles wide and dated at ninety million years.

The new Indian Ocean ice flow moving north scraped India to its bedrock and lifted up the Tibetan Plateau. It had traveled 5,500 miles in 55 million years[32]. The Ninety East Ridge, as seen today, points to the severest inland push of the ice flow.

The next illustration shows India floating, sliding and drifting a total of 5,500 miles in accordance with plate tectonics continental drift. Considering the Mid-Oceanic Ridge and plateau's placement, as well as the lack of any seafloor features supporting the theory that a land mass the size of India moved across this basin, plate tectonics should be set aside as an instrument of continent or mountain formation. The Himalayan Mountains were raised, along with the inland Tibetan plateau, by an ice flow, not the collision of tectonic plates.

If one were to push their foot in wet sand, a small bank will accumulate in the direction of motion, followed by a second and, if the force is maintained, a third as well, with each new bank being younger than the last. The bottom image illustrates this

---

[32] Six inches per year

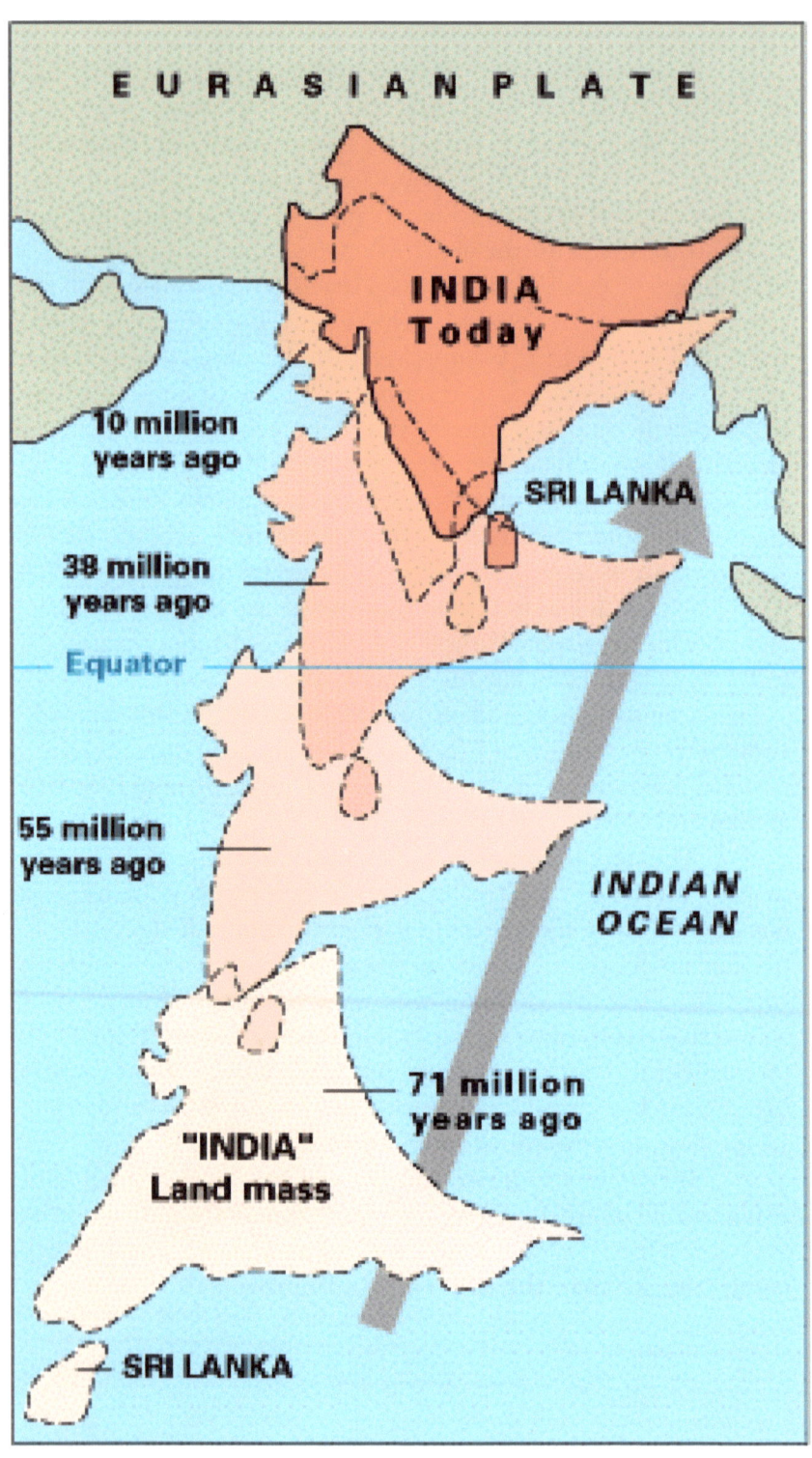

effect on a broader scale. The folds of Himalayan Mountains would eventually grow sufficiently to stop the massive ice flow.

The Himalayas were created over a 45 million year period from 55 Mya to 10 Mya. As in Russia, the Indian ice flow was stopped by the Tibetan plateau. After plowing out the Bay of Bengal east of India and the Arabian Sea to the west the flow was forced west, rifting and creating the arc of the Himalayas above India. This means the thin crust of the Indian Ocean was pushed north by the skimming action of an ice flow, rather than the mechanism of plate tectonics. If the latter had been responsible, the crust beneath the present-day Indian Ocean would be thick. Pausing at the knot of mountains called the Hindu Kush, the ice flow moved through what would be the Arabian Desert carving out the Red Sea - with its own "mid-ocean" ridge on the left and the Gulf of Oman to its right, simultaneously pushing up the folded mountains of Iran.

The ice flow that moved up the Red Sea was the same which excavated the Mediterranean Sea, moving over Cyprus, and Greece and up the peninsula of Italy. The flow thinned, and the progress towards the Alps was slow through the last 9 million years. Coral and limestone are present in the beautiful Dolomites of the Italian Alps. Limestone is present in the Croatian Fold

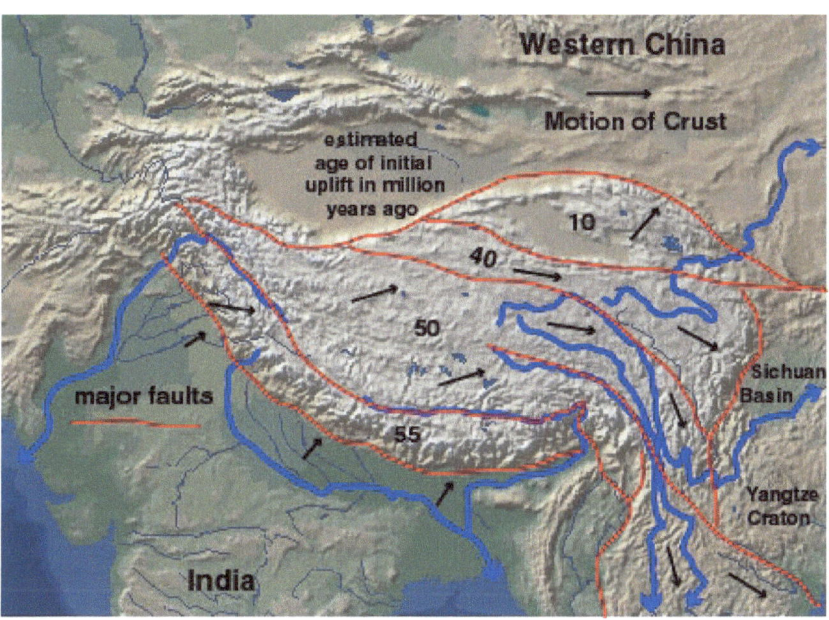

Mountains, which would have been on the eastern side of the flow. The Alps – that is, the French, Swiss, Italian, Austrian, and Croatian Alps - were pushed up from the south and folded over from the northwest in the form of a giant fan. Flying over the Swiss Alps, one can see waves of mountains heading north like those of a frozen ocean.

The Apennines, called the Spine of Italy, are volcanic. Croatia however is different: it has fold mountains like the Alps found elsewhere in Europe. When passing over this part of the world, the ice flow that lifted the bottom of the tropical sea not only formed limestone, but compressed it further into marble. The entire Apennine Mountain chain, then, also is an "ocean ridge" above sea level, and a rising landform as well, rebounding after being freed of the weight of the ice which formed it. The Italian Dolomites contain the remains of the same tropical sea which would form the Apennines. The frequent earthquakes from six miles beneath Italy's mountains are the signature of the 40,000 miles of ridges that cover the surface of the Earth.

Here's the evidence. 38,000 people have been killed in six earthquakes in the last century in Italy, and the entire ocean ridge earthquake profile matches the earthquake profile of the Apennine Mountains. Most earthquakes are six miles deep and most of the ridges are about 6000 feet high.

The presence of fold mountains to the north and east, and the volcanic geology of the mountains themselves are a classic ridge formation. The volcanoes remind us of the geological activity of the mid-ocean ridges, as well as those found in Iceland. A transcurrent fault even creates the "boot"-like shape of the Italian peninsula. This bears notable similarity to that of New Zealand, which is located diametrically opposite of Italy on the globe.

Take as a study the distribution of limestone in this region. The softer limestone is near Split, Croatia in the south, with harder marble to the north. The limestone here is so plentiful that the town's harbor district and plaza are paved with it. Split also bears the distinction of having supplied the marble for the White House of the United States. Italy's Michelangelo had to travel north of Rome to obtain the much harder "Carrara" marble.

When you view the mountains of Croatia compared to Italy you notice their structure differs considerably. In sharp contrast to this instability, there are no earthquakes for Corsica or the other large island of Sardinia.

Consider the island of Cyprus, which lies on the same path formed by the Apennines and ridges offshore through Greece. The geology of Cyprus was the ultimate cause of the cancellation of Project Mohole, the goal of which was to obtain a sample of the mantle by drilling into the Earth's crust where it was thinnest, at the ocean floor. It was discovered that the top of Cyprus had been scraped away, and mantle material exposed; indeed, Cyprus's caves are testament to the copper mining that persisted in that area for thousands of years. Copper "bleeds" from deeper molten rocks and settles in the hardened magma of the ocean ridge passing through the island.

Returning briefly to Sardinia, this island also contains copper ore, but is devoid of the basalt characteristic of an ocean ridge. This is as expected, as the island does not lie on the path of an ocean ridge to begin with. These facts, along with the absence of earthquakes and other qualities of the island's geology have lead scientists to refer to Sardinia as less an island and instead a small continent.

Given the unusual nature of these two islands, it is possible that Sardinia and Corsica are erratics that were pushed - or carried - by the ice flow and deposited in place when it melted approximately 9 million years ago.

Part 4

THE MECHANICS

We use latitude and longitude as a part of a coordinate system used to map locations on the Earth. Its principle unit of measure is the degree, defined as $1/360^{th}$ of a circle, and can be further divided into minutes ($1/60^{th}$ of a degree) and seconds ($1/60^{th}$ of a minute). The line equidistant from the north and south poles of the Earth is called the equator, and represents zero degrees latitude. Its longitudinal counterpart, the prime meridian, while not corresponding to any particular geographical feature, serves to represent zero degrees longitude. We use ninety degrees of both northerly and southerly latitude, beginning with zero degrees at the equator and ending with ninety degrees at each the north and south poles. We use one hundred and eighty degrees of both easterly and westerly longitude, with the prime meridian at zero degrees and the International Date Line at one hundred and eighty degrees east and west. In physical terms, one minute of latitude at the equator is equal to one nautical mile, and so one degree is equal to sixty nautical miles, which comes to sixty-nine statute miles. To bring this into broad perspective, this comes to a circumference of the Earth of 21,600 nautical or 24,900 statute miles.

As the poles are known to wander cyclically and historically, the equator also wanders in the same way. And as we can depict a circle of polar movement, we can also depict a circle of equator movement. Also keeping in mind that the magnetic north pole and the true north pole are not in the same location and do not have the same meaning, and that the magnetic poles movement varies more than the physical poles, we must be careful in our evaluation of magnetic striping in the crust of the Earth.

For the purpose of this text, dates are given in accordance with currently accepted scientific methods such as radiometric dating or carbon dating; it should be kept in mind that these systems are more reliable in determining relative ages than actual ages. Should straight line, non-linear, or logarithmic errors be discovered in these techniques in the future it could drastically change the actual dates but it would be unlikely to change relative placement in the geologic record.

With the knowledge of the speed and direction in which the ice flow was moving, we can locate where the equator was at any time in the last 248 million years. This is because the ice moved down the 30 degree west longitude line as indicated by the mid-Atlantic ridge, and up the 60 degree east longitude line as evidenced by the mid-oceanic ridge in the Indian Ocean. Keeping in mind that the spinning Earth would always push the ice toward the equator, this describes the path of Equator movement, with pivot points on the island of Hawaii and the area of Botswana, South Africa, the poles.

The mountains and thousands of fossils whose ages are well understood form strata which can be mapped to the Circle of Equator Movement. Due to axial precession,[33] the circle has a three degree variation amounting to a 180 nautical mile wandering of the equator.

Today's movement of the Earth's crust is different and of a lesser magnitude on account of the absence of massive ice deposits. The high continents of Antarctica and Asia are the new masses maintaining a four inch crustal movement toward the equator.

The geologic record contains many examples of coal in places it was not expected, like Svalbard-Spitsbergen in the Arctic Circle. The biodiversity of plants and animals that flourished in arctic regions can be explained by the six inch movement of the Earth's crust.

---

[33] Axial Precession is a slow, gravity induced change in the orientation of the axis of rotation an unbalanced spheroid.

"In the polar region there is a continued deposition of ice, which is not symmetrically distributed about the poles. The Earth's rotation acts on these asymmetrically deposited masses [of ice] and produces centrifugal momentum that is transmitted to the rigid crust of the earth. The constantly increasing centrifugal momentum produced in this way will, when it has reached a certain point, produce a movement of the Earth's crust over the rest of the Earth's body, and this will displace the Polar Regions toward the equator."

-Albert Einstein, from the forward to
The Path of the Pole by Charles Hapgood

I took the mid-point of all the Chandler wobbles which describe the circular motion of the Earth's axis of rotation. It was not like a spinning top; the crust of the planet was moving toward the west, over the axis of rotation. In modern times, this movement has slowed to just over four inches per year[34]. This deceleration is due to the decrease in unbalanced mass as the ice melted and an increase in momentum.

The movement is continuous and in one direction. Mountains were created by the massive asymmetrical ice cap causing crustal sliding and pushing up landforms, driven by the rotational forces of the Earth. The observation that this motion is in only one direction is congruent with the process of "true polar wander".

The movement started the ice flows 700 -800 million years ago in the northern Pacific Ocean, eventually causing the ice to spread over the rest of the planet. The equator 500 million years ago would have passed through modern Kansas in North America as indicated by quartz rock formations showing wave action. The movement down the Asian Continent indicates that the equator was in a line that is now southern Russia.

There is a connection between the moving ice shell, the crust sliding toward the equator, the 40,000 miles of mid-ocean ridges that begin in the Pacific region, the high fold mountains whose age decreases in correspondence with their position relative

---

[34] In 1976 the principal scientist of the Smithsonian Institute, after confirming my four inch calculation of the present day movement replied, "You are quite correct, and any further information you can provide would be quite helpful".

to the southerly and easterly progression of the Mid-Atlantic Ridge and the northerly movement of the Southwest Indian and Central Indian ridges. They are still rising, causing earthquakes six miles deep today.

The Earth is not a perfect sphere, and as a result different parts of its surface have differing moments of inertia. Despite these variances, the Earth's surface is, as is commonly reported, "Smoother than a billiard ball". However, when the massive ice cap was still situated over the Pacific, this was not so, and the variation would have been sufficient to destabilize the Earth's rotation. The ice mass on half the planet exerted a force pushing down on the crust and, in time, this would force the crust to rebound from the semi-plastic mantle below.

The rotation of the Earth was faster[35] when the Earth's water was solidified in ice. As a result the Earth had no tides. Furthermore, the moon may not as yet have had a significant presence in Earth's orbit, since at this point it still may have been within the Roche limit.[36]

The Earth is not a perfect sphere, but an oblate spheroid, which means the distance from the surface to the center decreases in proportion to its distance from the equator. Today, Earth's equatorial diameter is seventeen miles greater than its polar diameter because of its faster speed of rotation at the equator than at the poles. In early times when the planet had a faster spin the distance between the equatorial and polar diameters could have been much greater.

This rapid rotation would cause the huge mass of ice to impart its weight on the crust, and the rotational effects would

---

[35]J. Touma and J. Wisdom, 1958, Evolution of the Earth Moon System http://articles.adsabs.harvard.edu/cgi-bin/nph-iarticle_query?bibcode=1994AJ....108.1943T&db_key=AST&page_ind=15&data_type=GIF&type=SCREEN_VIEW&classic=YES: Adam Hadhazy | June 14, 2010 Fact or Fiction: The Days (and Nights) Are Getting Longer http://www.scientificamerican.com/article/earth-rotation-summer-solstice/

[36] The Roche limit is the minimum distance to which a large satellite can approach its primary body without being torn apart by tidal forces. If satellite and primary are of similar composition, the theoretical **limit** is about 2 1/2 times the radius of the larger body. *abyss.uoregon.edu/~js/glossary/roche_limit.html*

force the ice and top six miles of crust along the entire side of the Pacific hollow toward the equator. The massive amount of ice on half of the planet caused the planet to wobble, and in turn, the wobble caused the ice shell to move.

While its lowest extremity was seated six miles below the crust, the depth of the ice shell was far greater near the center than at its edges. Notably, the temperature of the earth at these depths is 527°F.[37]

The movement of the massive ice shell caused friction at the boundary with the crust, generating superheated steam. Such was the heat of this steam that it had the capacity to melt sold rock. The intuitive assumption would be that the ice would melt long before a temperature high enough to melt rock could be reached, but there are other factors at play. Given sufficient pressures, the temperature of steam can rise almost indefinitely, with machinery existing today which can produce temperature exceeding 1200°C, far above the 1000°C necessary to melt basaltic rock.[38]

The ice shell is a perfect heat exchanger; the heat is absorbed by the ice, which melts, turns to steam, rises, and refreezes again. It absorbed the heat of friction, and any heat rising from the core and it all was handled by the ice shell without melting it. Beneath the ice, pools of molten magma formed. Rock, unlike ice, is an "insulator"; it changes structure, retains the heat and only transmits this thermal energy by direct contact to the rock below.

Our planet has a solid core, a semi-plastic mantle and a crust. The mantle material beneath the crust is composed of a fictile basalt rock. While solid by conventional definition, on geological timescales the material of the mantle behaves as a viscous fluid; through the mechanism of convection, this ductile

---

[37] as calculated using the accepted geothermal gradient which suggests an average of a 75°F increase per mile underground.

[38] Engineers at the commercial boiler manufacturer Burnham Corporation informed me that given enough pressure you could raise the temperature as high as you like, high enough to melt rock.

rock would have been carried to the Pacific hollow and, over time, replaced some of the material lost in the moon forming collision.

Radioactive decay of material in the mantle would expand its volume, so that when it surfaced beneath the ice of the Pacific it would have driven it upward. The extent of this upheaval would have been significant, several miles at the edges and increasing near the center of the ice cap. Only a mass of ice of such proportions would be capable of pushing or plowing the crust material into the mountains we have today.

The ice shell prevented the mantle from returning to the Pacific. The leading edge pushed the crust material ahead of it and forced the viscous rock under the crust upward, thus raising it until it folded over on itself. This action caused the formation of layers in the crust preceding the ice.

The ice, miles deep and high also dragged the crust below with it. The push toward the equator would, over time, cause the entire crust to be divided into two parts, each rotating in the opposite direction while sliding over the mantle.

As mentioned in the introduction of this work, the wide lava flows which emerged from the trailing edge of the movement of the ice flows, in cooling and flowing anew, formed many steps, or layers, each around 45-60 feet in thickness. For example, the Columbia River plateau has 23 such steps, each of approximately this thickness. If you take Milankovitch's 413,000 year elliptical orbit and calculate the time it took to create all 23 steps, you arrive at 9.4 million years, an age quite close to that of the basalt of the lowest layer.

The Siberian steps of Russia, the Deccan traps of India and the steps in Greenland are all the same creation of the massive ice flow. Formed from the trailing edge of the Atlantic ice flow's counterclockwise path around the African continent, it took the same time to create a step in each location, about 1 million years each. It would have taken nearly 23 million years for the process to complete.

The energy required to push the crust against the forces generated by the Earth's more moderate circular orbits is already enormous; as the Earth's orbit increases in eccentricity, the ice in

Deccan Traps of India

turn would need to overcome a proportionally greater centrifugal force. As a result, the ice destabilized in the times of maximum eccentricity allowing the fold mountains that were forming to overcome the inertia of the great mass of ice, blocking forward movement and forcing the ice to move sideways.

This movement can be compared to a glacier which, sliding down the slope of a valley may be temporarily diverted around a large landform before returning to its prior, downward path, and can be seen in the mid-ocean ridges where the direction of the shift follows the continental contours.

The result is the "rifting" along the entire width of the advancing ice flow from the top to bottom. As was previously explained, the crust beneath the center of the ice shell would be subject to the greatest pressures and thus remain a long molten rift along the path of the ice: as the ice was diverted to the left and right, it would cut new, minor rifts perpendicular to the first. These elongated magma pools would eventually harden into transform or transcurrent faults.

One quality of these faults is their ability to record the orientations of earth's geomagnetic field. Since the magnetic pole was changing in relation to the motion of the crust, the alignment of magnetic particles in the molten rock would have changed as

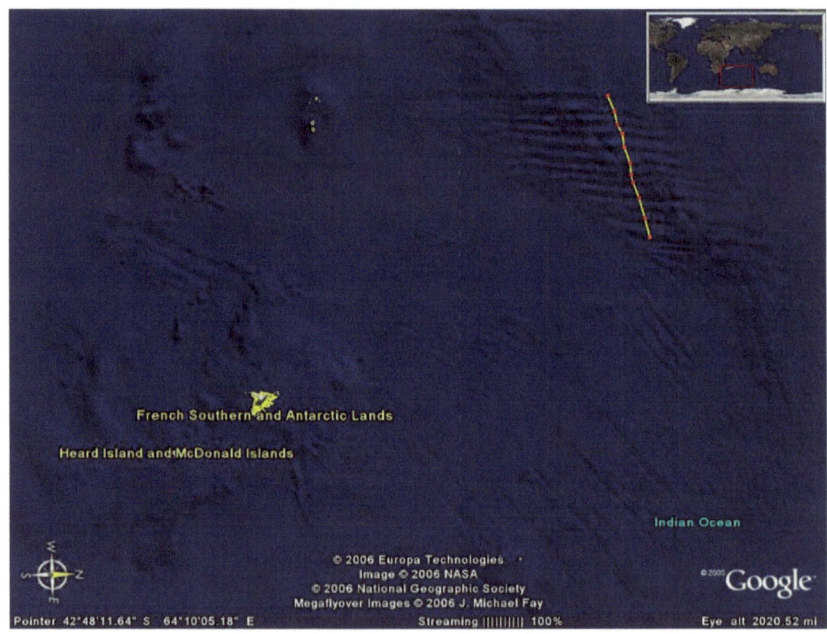

Transcurrent Faults on the Mid-Ocean Ridge

well. Once this basalt had cooled, these particles would no longer be free to change orientation. As a result, there is a visible pattern or footprint along these faults which serves as a record of the change of the position of the magnetic poles over time. This has been interpreted as proof of sea floor spreading, but this explanation is insufficient for a number of reasons.

For example, consider the distance between these ridges. If the ice was pushing the crust six inches per year, which is 100 miles per one million years, then according to the 413,000 year Milankovitch orbit cycle, distance between transform faults on average should be 41.3 miles. The image above shows transcurrent faults in the Indian Ocean. When measured, the faults average 41 miles apart, with less than a one percent degree of error.

Transform faults are not the only evidence of the correlation between the 413,000 year Milankovitch cycle and presence of a massive ice flow. Consider the aforementioned flood basalts; the same rifting which split the whole of the 3000 mile ice flow would cause a new layer of lava to be laid down over the old, adding another step every time Earth entered that unusual 413,000 year elliptical orbit.

Part 5

While the Ice Moved

It would be a mistake to assume the equator was always in the location it is today. The ice always flowed in the general direction from the poles to the equator, taking the most direct path where not diverted by masses of land. While the ice was flowing out from the northern and southern halves of the Pacific Basin, one of the earth's poles lay in the central Pacific. Since the ice flowed from the north Pacific over the Americas, then south through the Atlantic, around Africa and north in the Indian basin, it suggests that the equator was constantly shifting position over time. Despite the ice's considerable movement, there remained portions of the earth which were not frozen.

Strata indicating tidal change found in 900 million year old rocks prove that there was a liquid ocean on earth at that time, located in a narrow "habitable zone" parallel to the equator.[39] Interestingly, while parallel to the equator all species could wander throughout the zone. This habitable zone would slowly expand, but as the poles and equator shifted, ice melted, and this habitable zone moved, a tropical area developed in the area remaining equatorial, while temperate areas developed north and south of

---

[39] Revolutions that Made the Earth By Tim Lenton, Andrew Watson; 600-Million-Year-Old Embryos Found - By William J. Cromie, Harvard University Gazzette

that area. Once this occurred, the population of any species distributed through the zone could become divided: as a result, members of such a species could be found in the vicinity of both hemispheres without representatives found in the intervening tropical area. This is called "bi-polarity of species".

The fold mountains of the earth are like the rings in the trunk of a tree. Just as the rings record the number and character of the years as they pass, the mountains tell us where the ice was at a given period of time. The mid-ocean ridges provide a similar record. The timeline for the movement of the ice is equally important, and has major implications for the location of the Earth's mineral resources.

The publications of Charles Darwin and Alfred Russel Wallace in the mid-eighteenth century spawned an intensive examination of the geological record. Today we enjoy the fruits of those studies, by which we have gained tremendous insight into the nature of creatures both large and small. Likewise, our understanding of their habitats - largely areas of a temperate or tropical climate - can be traced through the 183 million years that they have lived. The slide of the earth's crust toward the changing equator can be traced through the fossil record of the dinosaurs across each of the continents... If one maps the cataloged dates and locations of these fossils and compares them to the hypothetical path of the ice flow, it is clear that they match up. Though the movement of the ice was slow – only six inches per year - the various species had to move a great distance to remain in the habitable zone as it followed the equator. Six inches a year is negligible compared with the fifteen hundred mile annual migration some species make today.

On all continents, different fossil sites have differing dates of formation. In North America, the fossils found in Texas are from the Cretaceous. Those found in Colorado or Wyoming are from the Jurassic, and if found from western Canada down to Montana, the fossils are from the Triassic period. This variation reflects the movement of the equator, following the Mid-Atlantic ridge during its counterclockwise movement around Africa.

During this time, the area of Botswana was the pivot point. The movement of the pole during the Mesozoic age (248 – 65 million years ago) is the easiest to trace. Using this general theory you can trace the location of dinosaur fossils to the habitable zone by using the mid-ocean ridges because the date at any site on the ridge indicates the location of the ice, and therefore the equator, at that time. The similarities of dinosaurs on all the continents indicate that the land area in front of and to the sides of the ice flow were a migratory path used by the dinosaurs as they moved from one place to another.

Plate Tectonics Theory explains the similarity of dinosaurs found one or more different continents via a different mechanism. It posits the movement of continents and formation of land bridges, allowing creatures to circumvent the enormous oceans of the earth

Fossils found in Kansas, South Dakota, and Wyoming show us that 82 million years ago, in the late Cretaceous period, there was an ocean which covered that area of the continent. Below is a fossil of an armored 8,000 pound fish native to that period. At this time, this inland ocean was draining into the Atlantic, freed on account of the withdrawal of the ice toward the equator by way of the Gulf of Mexico. This movement carried the sediments that would become the Yucatan Peninsula and Florida-Bahamas platform.

At the same time, the ice in the Pacific moved east, scouring out the basin which today is the Caribbean Sea. This was

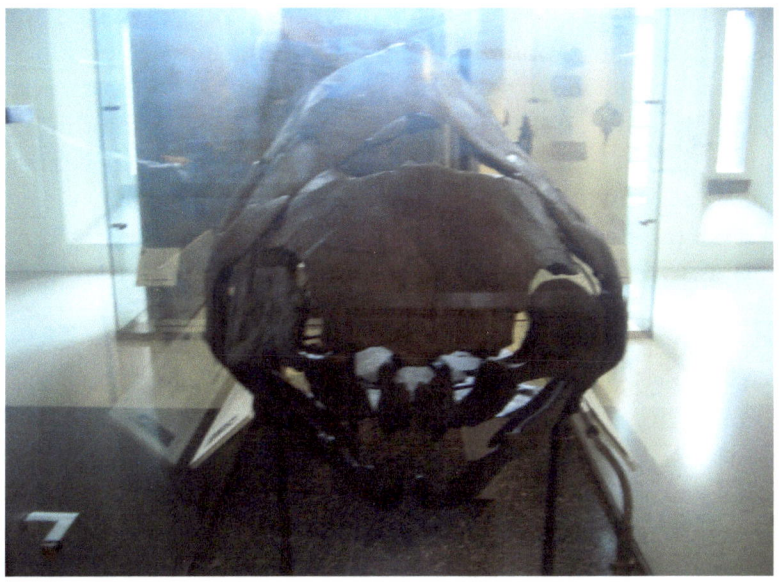

not the only basin created this way: compare it to the ocean floor just south of the Falkland Plateau off the coast of Tierra Del Feugo.

The Cantarell Oil Field in the Bay of Campeche leads me to believe deep oil should also be located under the Bahamas. Nature has a way of playing jokes on us, and I suspect oil will be found where oil is now being offloaded, as it is in the Bahamas at Freeport. Sea ice could be responsible for the accumulation of salt in this basin, by the same mechanism responsible for the decreased salinity of the Arctic Ocean. This process, called "fractional freezing", results in the desalinization of sea water, depositing salt in crystal form on the ocean floor. It is possible, over sufficient spans of time, that the layers of salt deposited in this way can reach thousands of feet in thickness. The thermal process at the base of the ice shell was also capable of contributing to the deposition of this salt.

As mentioned previously, there are two pivot points - Botswana South Africa and the big island of Hawaii – over which the ice shells would have turned. Eighty-two million years ago, the ice was sitting in the Gulf of Mexico, with the habitable zone a 900 foot deep inland sea that stretched from northern Canada to Mexico. Forests with well developed leaves and fronds, towering over thick underbrush of ferns, mosses, and liverworts, invaded the land. The massive coal deposits now found on the western side of the Appalachian Mountains were formed during this period.

A renegade ice flow would have overrun what may have been 20,000 to 30,000 foot high mountains pushed up by the passage of the primary ice flow. This flow would have left behind masses of ice which functioned as dams, allowing the inland sea to grow. With the movement of the equator, these would eventually fail. Just as when glacial ice dams fail in modern times, outburst floods carried sediment, plants, trees, insects, and animals more than 200 miles. The draining of the inland sea to the basins scraped out by ice flows was a catastrophic event, and created conditions resulting in the rapid burial of both animals and plants. This is the origin of many of the largest fossil beds we find today. The area previously occupied by the North American inland sea created the Smackover Formation of Arkansas and Louisiana and became a massive drainage basin with deltas emptying into the Gulf of Mexico. The fossils found in this basin are conclusive proof of the existence of the inland sea.[40]

There was a proliferation of species which began around 600 million years ago, and this can be traced to an expansion of the habitable zone, by that time covering nearly half of the earth. The fossils found in beach rocks that show the ancient tidal changes indicate water would have existed in the habitable zone for nearly 900 million years.

Alternating layers of organic debris and sand is called a cyclothem[41], or sequence stratigraphy. The area, a band about a thousand miles long and two-hundred miles wide on the west slope of the Appalachians, has seven distinctive coal seams within 2,000 feet of the surface.

The age of petroleum in this area is related to this drainage event. Petroleum engineers from the Bureau of Land Management speak of oil in the Jay Fields from the Jurassic Period in the panhandle region of Florida.[42] The inland sea of North America existed for hundreds of millions of years before the eighty-two million year old fossils mentioned above, and the sediments from it accumulated behind the northern wall of ice in the Gulf of Mexico, and are understood to be associated with the Jay Field oil.

The Destin Dome, south of Destin Florida, is an anticline structure[43] created after the ice scrapped the area to its bedrock. Through the application of the ice flow theory, it is clear that oil would not be found at this location. Despite spending billions of dollars[44] for the right to drill in this region, it was found dry. Around 125 million years ago, in the panhandle region to the west and specifically where Florida meets Alabama, there was a significant accumulation of sediment. In contrast to the Destin

---

[40] http://www.mnh.si.edu/exhibits/backyard-dinosaurs/finding-fossils.cfm

[41] Cyclothems are alternating stratigraphic sequences of marine and non-marine sediments, sometimes interbedded with coal seams.

[42] REASONABLY FORESEEABLE DEVELOPMENT SCENARIO FOR FLUID MINERALS, Prepared for: U.S. DEPARTMENT OF THE INTERIOR BUREAU OF LAND MANAGEMENT, EASTERN STATES, JACKSON FIELD OFFICE 411 Briarwood Drive, Suite 404, Jackson, MS 39206

[43] An *anticline* is a fold that is convex up and has its oldest beds at its core.

[44] http://archives.aapg.org/explorer/2001/10oct/destindome.cfm; Chevron/ Conoco

Dome, attempts to drill here found far greater success. More recently, I informed the four major oil companies that the Jay field is an accumulation area that extends east and west into deep water, six miles beneath the sand. South of that line, it is dry. The oil reserves in Jay field are likely being refilled from these deep oil locations.

Deep oil is less of a mystery when considered in the context of masses of ice extending miles deep into the crust. Deep oil cannot be explained any other way, unless it was in the ice and was buried, as an organic substance, by the ice shell movement. In my letter[45] to the oil giants, I informed them that the Desoto Canyon looks like a sediment structure, where a large ice mass allowed sediment to form around it as it melted leaving a canyon-like concavity. Canyon-like, as in contrast with a formation like the Grand Canyon, it lacks the imprint of rivers and streams at its edges.

In a part of Russia near the Yablonovy Range and Hentiyn Mountains, there is a geological curiosity called Lake Baikal. Aside from its unusual depth of five-thousand feet – giving the lake such volume as to hold twenty percent of all the fresh water on earth - it is home to a unique diversity of fauna, eighty percent of which are found nowhere else in the world. The reason for these unconventional characteristics is that it was at one point part of an ocean trench, created by the diversion of the Ural ice flow.

Ocean trenches occur when the ice shell flowed around immovable objects. Most of the original Russian ocean trenches that existed on the Russian platform have since been filled with sediment. What remains is Lake Baikal, which is slowly shrinking.

Many things are moving in unison over the Earth, most of them towards the equator. In 1996 Columbia University published research that indicated that the Earth's inner core rotates faster than its crust, making one total rotation inside the planet every 400 years. The instability of the magnetic currents of the molten inner core rotation causes geomagnetic drift and occasionally reversal. The rotation of the inner core also distributes the heat of in the outer core and mantle, which may conflict with the

---

[45] Letters were delivered to Exxon, British Petroleum, and PetroBras.

"hotspot theory" – the notion that plumes emanating from the Earth's core create large igneous provinces, or islands like as Hawaii, as the crust passes over them.

The Appalachian chain rose in sections; the Taconic 460 million years ago, the Acadian 390 to 250 million years ago and the Alleghenies 300 to 250 million years ago. In an article for *The American Naturalist,* dated July of 1877, American geologist, N.S. Shaler wrote:

> "..it would be nearer the truth to say that mountain systems are more likely the product of parallel upheavals appearing in successive geological periods than of single epochs of elevation."

In this quote, Shaler refers to the slow rise of mountains. This can be further seen in river systems that cut through the Appalachians. The rivers do not take the easy route around the mountains, as one might expect from the current lay of the land. These seemingly unintuitive paths are in truth records of earlier ages, when these rivers did indeed follow a path of least resistance downhill. Subsequent geologic forces changed the shape of the landscape, but these watercourses maintained a sufficient rate of erosion to match the uplift of the land around them. This is a near universal characteristic of the oldest rivers of the world.

Conclusion

This Ice Flow Theory of the Origin of Mountains is based on a new evaluation of the geologic record, taking into consideration the new data accumulated through GPS tracking, satellite imaging and the charting of the ocean floors. This is a new theory - not an advance in, addendum to, or modification of present theories though it takes into consideration the research of many ideas both present and past. It provides an explanation of the characteristics of the lithosphere above and below sea level and proposes the following:

1. An ice cap, approximately eighteen miles high, covered half of the planet one billion years ago in the area of the present day Pacific Ocean, whose weight caused it to press six miles into the lithosphere.

2. The ice cap destabilized the Earth's rotation, causing it to wobble.

3. The resultant wobble broke the ice cap loose, when, on account of the Coriolis Effect, it began to move in a clockwise direction around a pivot point in the Pacific.

4. The ice cap split near the Bering Sea.

5. The split divided the ice cap into two major flows, one remaining in the Pacific and the other flowing through the present-day Arctic towards Russia.

6. The ice flows traveled at a rate of six inches per year toward an equator that was shifting along with the crust over a viscous mantle.

7. The Coriolis Effect broke the Russian flow in two, one continuing toward Russia and the second flowing into the present day North Atlantic.

8. The Russian flow stopped at the Himalayas, turned clockwise and rejoined the Atlantic flow.

9. The Pacific flow was blocked by Antarctica and flowed west south of Australia.

10. The Atlantic flow stopped at Antarctica and turned around the southern extremity of Africa, where it joined the Pacific flow in the Indian Basin.

11. The combined flows continued north toward the Himalayas, then towards the Red Sea to the Mediterranean before coming to a stop at the Alps.

12. These ice flows scoured out the present day ocean basins leaving behind a weakened lithosphere where they were thickest, gouged out the ocean trenches, leaving cracks in the lithosphere and pushed up the Fold Mountain systems of today.

13. The part of the lithosphere weakened by the heaviest weight of the flows rebounded, creating the mid-ocean ridges.

Today we are in a period of late glacial rebound. Data from the Global Positioning System (GPS) shows motion in the vertical and horizontal that supports this. The GPS responders can even detect variations in elevation which would normally return false measurements of sea level. The Island of Great Britain, for example, is undergoing uplift in the north and subsidence in the south, so that at first brush it appears that sea level is falling in the north and rising in the south.

The Sichuan region in China is on the pressure ridge caused by the furthest push of the ice flow indicated by the Ninety Degree East Longitude Ridge. Records from sixty GPS stations in China indicate the Tibetan Plateau is moving north east while the Sichuan region is moving southeast. This region is being ripped apart at the boundary. The proof of this is the eastward flowing rivers of China[46].

---

[46] The Salween River, the Yangtze and the Mekong

The Sunda Trench, a fault in the northeastern Indian Ocean, lies on a path which passes through Sichuan. The crust on either side of this fault turns after the fashion of cog wheels, with each under the influence of a number of pivot points. Sichuan is being influenced by the stress found midway between two pivot points, near the Hawaiian Islands and Botswana respectively. Along with Sichuan, northern Sumatra, the site of the disastrous 2004 Indian Ocean earthquake, sits at the midpoint between these sites. This earthquake was the result of the massive pressures on either side of the Sunda Trench. With this knowledge, the trench walls can be monitored, so as to detect signs of such geological instability in the future. Similar to the breaking of submarine cables off the coast of Hokkaido before the 2003 Tokachi-Oki Earthquake, there are many events which can offer some degree of forewarning of impending seismic activity.

The Zipingpu Reservoir in China, filled in 2006 for the first time, could have caused China's Sichuan earthquake. The Zipingpu Reservoir holds 1.2 billion cubic meters of water, a weight more than four times that of the seven billion people living on Earth.[47] Adding this much weight in an unstable area can have catastrophic effects. This reservoir is only three miles from the epicenter and experienced 12,000 aftershocks the month following the quake.

---

[47] a - Weight of all humans: ~632 billion lbs; b - Weight of 1.2 billion cubic meters of water: ~2.65 trillion lbs; b/a = 4.184

Afterword

This book, The Origin of Mountains, has a potential use in the identification of the location of oil and coal deposits. By following the timeline of the mid-ocean ridges, the location of the ice at any given time can be determined, and with that also the location of the habitable zone, and thus the location of oil and coal. However, it is unlikely that oil will be found anywhere the lithosphere was scraped away by the movements of the ice; it is more likely to be found in regions unscathed like the Kerguelen plateau, or in drainage basins away from the fold mountains.

Appendix

List of Illustrations

The Eiger, a 13,000 foot dislocated mountain	Cover
Patagonia, Argentina – Peaks 100 MYA, Plateau 26 Mya	6
Theoretical Timeline of the Earth	8
Circle of Prolific Oil Wells	13
GEOSAT	16
Large Igneous Provinces – Flood Basalts	20
Pacific Ocean	23
Footprints of Ice Movement	31
The Development of Mid-Ocean Ridges	34
The Lewis Overthrust, Proterozoic Rock over Cretaceous	36
Ice Cap Breakup and Movement	38
Age of the Mid-Atlantic Ridge	43
Continental Drift Theory vs. GPS Tracking	44
Circular Formations on the Karoo Large Igneous Province	47
Dating along the Sunda Trench	50
Movement of India via Continental Drift Theory	52
Dating the Himalayan Plateau	53
Deccan Traps of India	63
Transcurrent Faults on the Mid-Ocean Ridge	64
Armored Fish from Cretaceous Period	67

List of Flood Basalts

1. Chilcotin Group (south-central British Columbia, Canada)
2. High Arctic Large Igneous Province
3. Columbia-Snake River flood basalts (see Columbia River Basalt Group)
4. Ethiopian and Yemen traps in the Ethiopian Highlands
5. Viluy traps
6. Pre-Devonian traps
7. North Atlantic Volcanic Province
8. Emeishan Traps (western China)
9. Deccan Traps (India) 35 million years ago (end of Cretaceous period)
10. Caribbean large igneous province
11. Mackenzie Large Igneous Province
12. Kerguelen Plateau
13. Ontong Java–Manihiki–Hikurangi Plateau
14. Paraná and Etendeka traps (Brazil-Namibia)
15. Karoo and Ferrar provinces (South Africa-Antarctica)
16. Central Atlantic Magmatic Province
17. Siberian Traps (Russia) 248 million years ago (end of Permian)

I thus learnt my first great lesson in the inquiry into these obscure fields of knowledge, never to accept the disbelief of great men or their accusations of imposture or of imbecility, as of any weight when opposed to the repeated observation of facts by other men, admittedly sane and honest. The whole history of science shows us that whenever the educated and scientific men of any age have denied the facts of other investigators on a priori grounds of absurdity or impossibility, the deniers have always been wrong.

Alfred Russel Wallace, 1893

Other References:

Delano's Discovery - A Hearthstone book Carlton Press, Inc- New York, N. Y. ISBN 0-8062-2696-x Copyright 1986.

Episodes of Flood-Basalt Volcanism Defined by Ar/Ar Age Distributions: Correlation with Mass extinctions? Haggerty, Journal of Undergraduate Science 3: 155-164 (Fall 1996) NYU

Ivanhoe, L. F, and G G. Leckie, -"Global oil, gas fields, sizes tallied, analyzed," Oil and Gas Journal. Feb. 15, 1993, pp. 87-91

Li Guoyu (2011), World Atlas of Oil and Gas Basins (Oxford: Wiley-Blackwell), p. 20.

May 2015
John Delano
38 Old Sylvan Lake Road
Hopewell Jct., NY 12533
Sunyday1@optonline.net

www.ingramcontent.com/pod-product-compliance
Lightning Source LLC
Chambersburg PA
CBHW041103180526
45172CB00001B/82